T0201320

DNS SECURITY
MANAGEMENT

DNS SECURITY
MANAGEMENT

Michael Dooley
Timothy Rooney

IEEE Press
Series on
Networks and
Services Management

Thomas Plevyak and
Veli Sahin, *Series Editors*

IEEE PRESS

WILEY

Published by John Wiley & Sons, Inc., Hoboken, New Jersey.
Published simultaneously in Canada.

For general information on our other products and services or for technical support, please contact our Customer Care Department within the United States at (800) 762-2974, outside the United States at (317) 572-3993 or fax (317) 572-4002.

Wiley also publishes its books in a variety of electronic formats. Some content that appears in print may not be available in electronic formats. For more information about Wiley products, visit our web site at www.wiley.com.

Library of Congress Cataloging-in-Publication Data is available.

ISBN: 978-1-119-32827-8

Printed in the United States of America.

10 9 8 7 6 5 4 3 2 1

CONTENTS

PREFACE

As the Internet transformed from a scientific and educational network experiment into a global commercial communications network over the last four decades, its widespread adoption by organizations and individuals fueled its explosive growth, which continues unabated to this day. Providing the ability to communicate, update bank accounts, access critical information remotely, and much more, the Internet serves as the prime vehicle for anytime, anyway, anywhere communications. As such it also became an extremely attractive target for criminals seeking to disrupt an organization's web presence or email, infiltrate an enterprise's internal network, and steal valuable data, including personal, corporate, or government classified information, among other things.

Organizations have responded by implementing various security measures including intrusion protection mechanisms such as firewalls, authentication and encryption technologies, proactive security scanning, attack detection monitors, and security education to name a few. With defensive implementations deployed, attackers seek targets that are less well defended or that provide a generally free-flowing communications pathway like hypertext transfer protocol (HTTP), simple mail transfer protocol (SMTP), or domain name services (DNS).

While attacks on web traffic and email have been fairly well publicized, though they continue to evolve over time, attacks on DNS represent an emerging threat to organizations. Unlike HTTP, SNMP, and other application layer protocols, DNS is an "application helper" protocol that exists to facilitate application ease-of-use. DNS is not an end user application, but without DNS end user applications would largely be useless.

DNS essentially serves as the Internet directory to convert user-entered web or email addresses that humans use into Internet Protocol (IP) addresses computers use. And with each web page to which one navigates DNS provides analogous lookup functions for additional content all of which is "linked" from the destination web page, including images, scripts, styles, videos, ads, and so on. DNS also provides lookup functions for numerous other applications such as machine-to-machine communications, the Internet of Things (IoT), voice over IP, and so on. DNS is truly indispensable to the effective operation of the Internet and of your IP network.

The objective of this book is to help you understand how DNS works, its vulnerabilities, threats and attack vectors, and how to incorporate detection, defensive, and mitigation techniques to secure your DNS. By securing your DNS, you can better secure your network at large. We've attempted to bring together the topics of DNS and security assuming only basic knowledge of both. In the first chapter, we set the stage with a brief introduction to DNS and the National Institute of Standards and Technologies (NIST) Cybersecurity Framework (1), which is a de facto security implementation standard not only for the US government, but for organizations worldwide. This framework defines a common lexicon to facilitate documentation and communication of security requirements and level of implementation. In addition, the framework enables an organization to identify risks and to prioritize the mitigation of risks with respect to business priorities and available resources.

Chapters 2 and 3 provide an introduction to DNS, with Chapter 2 covering the organization of DNS data across the Internet and Chapter 3 discussing the DNS protocol which serves as a foundation for understanding certain attack methods. Chapter 4 introduces these attack methods at a summary level. The purpose of this chapter is to cover the breadth of attack types, which we've segmented into those that attack DNS itself and those that use DNS as an attack enabler to target other computing and network systems.

Chapters 5 through 11 serve as the heart of the book, drilling into each of the vulnerabilities introduced in Chapter 4 and providing detailed discussions of each attack form, detection techniques, defensive measures, and mitigation methods. We apply the relevant cybersecurity framework principles throughout these chapters to normalize the discussion within a security context. As we discuss various protection and mitigation strategies, we've also included basic configuration syntax to implement these strategies for three popular open source DNS implementations: Internet Systems Consortium (ISC), PowerDNS, and Knot DNS. Note that there are dozens of DNS server implementations and we chose these based on their free availability and relatively rich features sets, but we invite you to evaluate other implementations that may suit your particular requirements.

Chapter 12 brings things together in discussing an overall DNS security strategy. And once you've secured your DNS, you can then use DNS as a means to enhance the security of your overall network and applications. In Chapter 13, we discuss methods for using DNS to secure critical applications including web browsing, email, and others. Chapter 14 provides a brief though bold prediction of our view of the evolution of DNS security techniques and technologies. We've also included a pair of appendices, the first mapping the NIST framework core to DNS-specific outcomes and the second providing a list of currently defined DNS resource records, which summarily illustrate the diversity of data type lookups that DNS offers.

These network or cyber security technologies must continue to evolve to protect against a growing volume and diversity of attacks. As new defensive measures are implemented, criminals deftly seek new and innovative methods to attack targets to suit their nefarious purposes. Attack strategies have grown increasingly sophisticated as have defense mechanisms in response. As this arms race spirals onward, don't lose sight of protecting your DNS. We hope this book helps you understand the threats against your DNS and the mechanisms you can implement to defend and protect against them.

MICHAEL DOOLEY
TIMOTHY ROONEY

ACKNOWLEDGMENTS

We would both like to thank Thomas Plevyak, Veli Sahin, and Mary Hatcher, our editors at IEEE Press for their ongoing support and encouragement. We'd also like to thank Scott Rose, Paul Mockapetris, Paul Vixie, Greg Rabil, and Stu Jacobs for their time spent reviewing proposals and drafts of this book and for providing extremely valuable feedback. Their feedback has vastly improved the quality and usefulness of this book.

Michael: I would like to thank my family, my wife Suzanne, my son Michael, and my daughter Kelly, for all their love and support and allowing me to be distracted at home while I was working on this book. And I can't forget my lovable dog Bailey and crazy Ollie as well. I would also like to thank the following individuals who are my friends and co-workers. I have had the pleasure to work with some of the best and brightest people in the world, and I am truly blessed. In no particular order: Karen Pell, Steve Thompson, Greg Rabil, John Ramkawsky, Alex Drescher, and Bob Lieber. I would also like to acknowledge the original Quadritek leadership team that I had the privilege to work with as we helped to define and create the IP Address Management market back in the early years, specifically including Joe D'Andrea, Arun Kapur, and Keith Larson.

Timothy: I would first like to thank my family especially my wife LeeAnn, and my daughters Maeve and Tess as well as Uncle Jimmy for their love and support during the development of this book. I would also like to thank the following individuals with whom I have had the pleasure to work and from whom I have learned tremendously about networking technologies among other things: Greg Rabil, John Ramkawsky, Andy D'Ambrosio, Alex Drescher, Steve Thompson, David Cross, Marco Mecarelli, Brian Hart, Frank Jennings, and those I have worked with at BT, Diamond IP, INS, and Lucent. From my formative time in the field of networking at Bell Laboratories, I thank John Marciszewski, Anthony Longhitano, Sampath Ramaswami, Maryclaire Brescia, Krishna Murti, Gaston Arredondo, Robert Schoenweisner, Tom Walker, Charlene Paull, Frank DeAngelis, Ray Pennotti, and particularly my mentor, Thomas Chu. I also wish to acknowledge others who have otherwise inspired me to press on to complete this and affiliated works, including Peter Tsai, Elle Carpenter, Howard Falick, Holly Weller, Steve Wheeler, Ken Schumaker, Martin Wellsted, Craig Hamilton, and my esteemed co-author, Michael Dooley.

1

INTRODUCTION

WHY ATTACK DNS?

The Domain Name System (DNS) is fundamental to the proper operation of virtually all Internet Protocol (IP) network applications, from web browsing to email, multimedia applications, and more. Every time you type a web address, send an email or access an IP application, you use DNS. DNS provides the lookup service to translate the website name you entered, for example, to its corresponding IP address that your computer needs to communicate via the Internet.

This lookup service is more commonly referred to as a *name resolution* process, whereby a worldwide web "www" address is resolved to its IP address. And a given web page may require several DNS lookups. If you view the source of a random web page, for example, count the number of link, hypertext reference (href), and source (src) tags that contain a unique domain name. Each of these stimulate your browser to perform a DNS lookup to fetch the referenced image, file or script, and perhaps pre-fetch links. And each time you click a link to navigate to a new page, the process repeats with successive DNS lookups required to fully render the destination page.

Email too relies on DNS for email delivery, enabling you to send email using the familiar user@destination syntax, where DNS identifies the destination's IP address for transmission of the email. And DNS goes well beyond web or email address resolution. Virtually every application on your computer, tablet, smartphone, security

DNS Security Management, First Edition. Michael Dooley and Timothy Rooney.
© 2017 by The Institute of Electrical and Electronic Engineers, Inc. Published 2017 by John Wiley & Sons, Inc.

cameras, thermostats, and other "things" that access the Internet require DNS for proper operation. Without DNS, navigating and accessing Internet applications would be all but impossible.

Network Disruption

An outage or an attack that renders the DNS service unavailable or which manipulates the integrity of the data contained within DNS can effectively bring a network down from an end user perspective. Even if network connectivity exists, unless you already know the IP address of the site to which you'd like to connect and enter it into the browser address field, you'll be unable to connect, and you won't see any linked images or content.

Such an event of the unavailability of DNS will likely spur a flurry of old fashioned phone calls to your support desk or call center to politely report the problem. IP network administrators generally desire to minimize such calls to the support center, polite or otherwise, given that it forces those supporting the network to drop what they're doing and resolve the issue with the added pressure of visibility across the wider IT or Operations organization.

DNS as a Backdoor

Just as DNS is the first step in allowing users to connect to websites, it is likewise usable by bad actors to connect to internal targets within your enterprise and external command and control centers for updates and directives to perform nefarious tasks. Given the necessity of DNS, DNS traffic is generally permitted to flow freely through networks, exposing networks to attacks that leverage this freedom of communications for lookups or for tunneling of data out of the organization.

Thus, attacking DNS could not only effectively bring down a network from users' perspectives, leveraging DNS could enable attackers to communicate to malware-infected devices within the network to initiate internal attacks, to exfiltrate sensitive information, or to perform other malicious activity. Malware-infected devices may be enlisted to serve as remote robots or *bots* under the control of an attacker. A collection of such bots is referred to as a *botnet*. A botnet enables an attacker to enlist an army of devices potentially installed around the world to perform software programmable actions.

By its very nature, the global Internet DNS system serves as a distributed data repository containing domain names (e.g., for websites) and corresponding IP address information. The distributed nature of DNS applies not only to the global geographic distribution of DNS servers, but to the distribution of administration of the information published within respective domains of this repository. DNS has proven extremely effective and scalable in practice and most people take DNS for granted given this and its historical reliability. However, its essential function and decentralized architecture serve to attract attackers seeking to exploit the architecture and rich data store for sinister activities.

While DNS is the first step in IP communications, many enterprise security strategies trivialize or startlingly even ignore its role in communications and therefore its susceptibility to attacks on this vital network service or on the network itself. Most security strategies and solutions focus on filtering "in-band" communication flow in order to detect and mitigate cyber attacks. However, as we shall see, filtering DNS traffic can support a broader network security plan in providing additional information for use in identifying and troubleshooting attack incidents. This book is intended to provide details regarding the criticality of DNS, its vulnerabilities, and strategies you can implement to better secure your DNS infrastructure, which will in turn better secure your overall network.

DNS BASIC OPERATION

Figure 1.1 illustrates the basic flow of a DNS query. Upon entry of the desired destination by name, www.example.com in this case, software called a *resolver* is invoked by the application, for example, web browser. This resolver software is typically included with the device operating system. If a connection had recently been made to this website, its IP address may already be stored in the *resolver cache*. The resolver cache helps improve resolution performance by temporarily keeping track of recently resolved name-to-IP address mappings. In such a case, the resolver may return the IP address immediately to the application to establish a connection without having to query a DNS server.

If no relevant information exists in the resolver cache the device will query its *recursive DNS server*. The role of the recursive server is to locate the answer to the

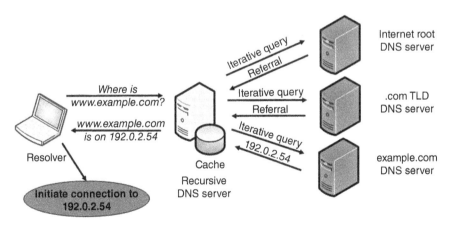

Figure 1.1.   Basic DNS Resolution Flow

device's query. The recursive server is itself a resolver of the DNS query; we refer to the resolver on the originating device as a *stub resolver* as it initiates a query to its recursive server, and it relies solely on the recursive server to locate and return the answer. The stub resolver is configured with DNS server IP addresses to query as part of the IP network initialization process. For example, when a device boots up, it typically requests an IP address from a dynamic host configuration protocol (DHCP) server. The DHCP server can be configured to not only provide an IP address but the IP addresses of recursive DNS servers to which DNS queries should be directed. Use of DHCP in this manner facilitates mobility and efficiency as addresses can be shared and can be assigned based on the relevant point of connection to the IP network.

As we mentioned, the recursive DNS server's role is to resolve the query on behalf of the stub resolver. It performs this role using its own cache of previously resolved queries or by querying DNS servers on the Internet. The process of querying Internet DNS servers seeks to first locate a DNS server that is *authoritative* for the domain for which the query relates (example.com in this case) and then to query an authoritative server itself to obtain an answer that can be passed back to the client, thereby completing the resolution process. The location of the authoritative server is determined by querying Internet DNS servers that are responsible for the layers of the domain tree "above" or "to the right" of the domain in question. We'll discuss this process in more detail in Chapter 2. The recursive server caches the resolution information in order to respond more quickly to a similar query without having to re-seek the answer on the Internet.

To access your website, people need to know your web address, or technically your uniform resource locator or URL. And you need to publish this web address in DNS in the form of a *resource record* so browsers can locate your DNS servers and resolve your www address to your web server's IP address. Multiple, at least two, authoritative DNS servers must be deployed to provide services continuity in the event of a server outage. Generally, an administrator configures a *master* server that then replicates or transfers its domain information to one or more *slave* servers. We will discuss more details on this process and server roles in Chapter 2.

Basic DNS Data Sources and Flows

Figure 1.2 illustrates a subset of the various data stores for DNS data and corresponding data sources. The authoritative DNS servers must be configured to answer queries for domain name-to-IP address mappings for this domain for which they are authoritative. Depending on your DNS server vendor implementation, DNS configuration information may be supplied by editing text files, using a vendor graphical user interface (GUI) or deploying files from an IP address management (IPAM) system as shown in Figure 1.2. Each server generally relies on a configuration file and authoritative servers store DNS resolution information in zone files or a database. Some implementations utilize dynamic journal files to temporarily store DNS information

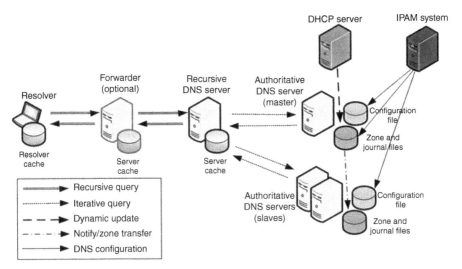

Figure 1.2.   DNS Query Flow and Data Sources

updates prior to committing to zone files in an effort to improve performance. All vendor implementations feature the ability to update DNS resolution information on a given "master" server which will then replicate this information to other authoritative servers to provide redundancy of this information.

Figure 1.2 also illustrates various DNS message types that are used to query for and configure DNS information. The recursive and iterative query types enable the resolution of DNS data, while dynamic updates and zone transfers enable the dynamic updating and replication of resolution information, respectively. DNS configuration information includes the parameters of operation for the DNS server daemon as well as published resolution data within zone files. We'll describe these in more detail in Chapters 2 and 3, but for now, you may observe that there are several independent data sources that may configure your DNS information as you permit.

DNS Trust Model

The DNS trust model refers to how DNS information flows among these components of the DNS system. In general, information received by other components in the system is trusted though various forms of validation and authentication can improve trustworthiness as we shall discuss later.

From the client resolver perspective, the client trusts its resolver cache and the recursive server to provide answers to DNS queries. Should either trusted data source be corrupted, the resolver could inadvertently redirect the user application to an inappropriate destination. For example, a user, thinking he or she is connecting to his or her bank, may inadvertently be connected to an imposter site in an attempt by an attacker to collect authentication credentials or financial information.

The recursive server trusts its cache and the various DNS servers it queries, whether internally on the enterprise network or externally on the Internet. It relies not only on accurate responses from authoritative DNS servers but on other domain servers which provide referrals to locate DNS servers authoritative for the domain in question. Referral answers are generally provided by the Internet root servers as well as top level domain ("TLD," e.g., .com, .edu, .net, etc.) servers as shown in Figure 1.1. Referrals may also be provided by other servers operated internally, externally, or by DNS hosting providers to walk down the hierarchical domain tree to locate the authoritative DNS servers. Corruption of recursive server DNS information, whether referral or resolution data, could have broader impacts affecting many clients given the caching of seemingly accurate resolution data. Each user attempting to connect to a website whose resolution information has been poisoned may be provided such falsified information from the recursive server cache.

Authoritative DNS servers are so called given that they are purportedly operated by or on behalf of the operator of a given DNS domain who is responsible for the information published on these servers. Resolvers attempting to resolve hostnames within the domain of the authoritative server trust the server to respond with accurate information, where *accurate* means *as published by the domain administrator*. Information published within authoritative DNS servers originates from a variety of sources as shown in Figure 1.2, including manually edited text files, inter-server transfers and updates, and/or use of IPAM solutions. Inter-server transfers refer to master–slave replication, while updates may originate from other DNS servers, DHCP servers, other systems, or even end user devices if permitted by administrators. Corruption of authoritative server configuration information impacts all Internet users attempting to connect with resources within the corresponding domain.

DNS Administrator Scope

As a DNS administrator, you'll generally need to be concerned first with your internal users or customers attempting to resolve domain names within your internal infrastructure and those on the Internet. For access to your internal systems, you'll need to configure authoritative DNS servers with the domain name to IP address mappings for those systems for which internal users need access. We'll refer to this naming and address mappings for internal infrastructure as your *internal namespace*.

To enable your users to access Internet websites, you'll need to manage recursive servers which your users can query to locate external authoritative servers from which to seek query answers.

You will also need to provide external Internet and extranet users with name to address mappings for your Internet reachable systems such as websites and email servers. Note that this *external namespace* will likely be a subset of your internal namespace, though ideally totally independent. We'll discuss approaches to serving these constituencies beginning with respective DNS server deployment approaches in Chapter 5.

SECURITY CONTEXT AND OVERVIEW

The practice of network security essentially boils down to the management of risks against a network. Risks may consist not only of malicious attacks but also include natural or man-made disasters, poor architecture design, unintended side effects of legitimate actions, and user error. Development of a security plan requires enumeration of risks, identification of vulnerabilities which may be exploited to affect the risk, characterization of the likelihood of each risk, determination of the impact the risk presents to the organization and defining controls to constrain the risk impact for each. Application of controls seeks to mitigate the risk to eliminate, or more likely yield a lower level of residual risk that is more tolerable to the organization.

We will apply the National Institute of Standards and Technologies (NIST) Cybersecurity Framework (1) as the context within which we discuss security approaches and strategies. The cybersecurity framework has emerged as a de facto standard worldwide. While no security framework can be "one size fits all," it provides a taxonomy and methodology for organizations to characterize their current and desired (planned) cybersecurity status, to prioritize initiatives to enable improvement of their current status toward their desired state and to communicate among stakeholders about cybersecurity risk. The framework provides guidance for organizations to perform risk assessments and to plan to manage risk in light of each individual organization's vulnerabilities, threats, and risk tolerance.

The framework relies on existing security standards including COBIT 5 (2), ISA 62443 (3), ISO/IEC 27000 (4), NIST SP 800-53 Rev4 (5) among others. It references specific sections of these supporting standards within each of the major framework activities. As such, the framework essentially provides a common overlay among these various standards to define a language for expressing and managing cybersecurity risk.

Cybersecurity Framework Overview

NIST's cybersecurity framework seeks to facilitate communications within and external to an organization when conveying security goals, maturity status, improvement plans, and risks. The framework is comprised of three major components:

- The *framework core* defines security activities and desired outcomes for the lifecycle of an organization's management of cybersecurity risk. The core includes detailed references to existing standards to enable common cross-standard categorization of activities. The core defines these activities across five functions.

o Identify – deals with what systems, assets, data, and capabilities require protection

o Protect – implement safeguards to limit the impact of a security event

o Detect – identification of incidents

o Respond – deals with security event management, containing incident impacts

o Recover – resilience and restoration capabilities

Each function has a set of defined categories and subcategories which we will explore later in this chapter.

- The *framework profile* defines the mechanism for assessing and communicating the current level of security implementation as well as the desired or planned level of implementation. The profile applies business constraints and priorities, as well as risk tolerance to the framework core functions to characterize a particular implementation scenario.

- The *framework implementation tiers* define four gradations of maturity level of security implementations, ranging from informal and reactive to proactive, agile and communicative.

o Tier 1 – Partial – Informal, ad hoc, reactive risk management practices with limited organizational level risk awareness and little to no external participation with other entities.

o Tier 2 – Risk Informed – Management approved with widely established organization-wide risk awareness but with informal and limited organization-wide risk management practices and informal external participation.

o Tier 3 – Repeatable – Risk management practices are formally approved as policy with defined processes and procedures which are regularly updated based on changes in business requirements as well as the threat and technology landscape. Personnel are trained and the organization collaborates with external partners in response to events.

o Tier 4 – Adaptive – Organization-wide approach to managing cybersecurity risk where practices are adapted to the changing cybersecurity landscape in a timely manner. The organization manages risk and shares information with partners.

The implementation tiers enable an organization to apply the rigor of a selected maturity level to their target profile definition to align risk management practices to the particular organization's security practices, threat environment, regulatory requirements, business objectives, and organizational constraints.

Framework Implementation

Implementation of the cybersecurity framework entails interaction and feedback among three major organizational tiers.

- Executive level – with a focus on organizational and business risk, the executive level sets out business priorities, risk tolerance, and security budget to those in the business/process level.
- Business/Process level – in consideration of business priorities, risk tolerance, and budget, this level focuses on critical infrastructure risk management, defining a cybersecurity framework target profile for the organization based on these inputs and the current profile, allocating budget accordingly to closing gaps between these profiles. This level feeds back to the Executive level any changes in current and future risk based on security threats and technologies and provides implementation directives to the Implementation and Operations level.
- Implementation/Operations level – responsible for framework profile implementation and risk management tactics. Feedback to the business/process level includes implementation progress, issues, and changes in assets, vulnerabilities, and threats.

The cybersecurity framework document leverages this three-tier organizational structure and identifies the following basic steps in defining a cybersecurity plan:

1. The first step starts at the Executive level to identify your business and organizational priorities and objectives and risk tolerance in order to scope out in priority order the set of assets and systems within the network to focus on.
2. Within the selected scope, the second step entails the organization enumerating affected systems and assets, regulatory and legal requirements, risk tolerance, and corresponding threats and vulnerabilities associated with the scoped systems and assets.
3. This step consists of defining the current status of cybersecurity implementation. Using the framework core, you can identify your level of compliance and discipline in implementing each function category and subcategory. The resulting analysis becomes your Current Profile defining a snapshot of your organization's alignment with the framework.
4. A risk assessment should then be conducted to enumerate risks in terms of asset vulnerabilities, potential threats and respective likelihood, and the potential network and business impact of each threat.
5. The fifth step entails defining the desired cybersecurity activities and outcomes by defining the Target Profile. Using the framework core along with business-specific categories and subcategories, desired outcomes can be enumerated.

6. Comparing the Target Profile with the Current Profile, one may define the gaps which need to be addressed to evolve from the current status to the desired state. Based on the cost to implement gap closure for each category and sub-category in light of its corresponding security priority, the business can determine whether to invest in closing that gap based on the corresponding value to the business. This helps prioritize which gaps will be addressed initially, which can be addressed later, and at what cost for each from a capital, expense, and resource perspective.

7. The final step consists of formally defining and implementing an action plan to address the prioritized gaps. As implementation ensues within the Implementation/Operations level, and snapshots of current or in-progress status may be communicated by updating the Current Profile.

The Current and Target profiles enable communication within or outside an organization of its current and planned cybersecurity implementation state, respectively. The broad use of this common framework facilitates communication among these entities and stakeholders using well-defined terms.

Scoping DNS   Once your executive team identifies and prioritizes DNS as within the scope of priority for applying security controls, your business team needs to define the corresponding set of affected DNS components. Table 1.1 illustrates an example scoping of basic business priorities that affect DNS to corresponding affected DNS components which could be considered for application of security controls.

TABLE 1.1   DNS Scope Examples

Broad DNS Scope	Affected DNS Components
Accurately resolving the organization's published namespace on the Internet	• Authoritative DNS servers and/or your external DNS hosting provider configured to resolve your namespace
Accurately resolving the organization's namespace for internal users	• Device stub resolvers • Recursive DNS servers configured to resolve DNS queries from internal device resolvers • Authoritative DNS servers configured to resolve your namespace for internal resolvers
Accurately resolving Internet domain names for legitimate internal user access to the Internet	• Device stub resolvers • Recursive DNS servers configured to resolve DNS queries from internal device resolvers

Current Profile Once the organization defines the scope as including one or all of these broad areas, a *Current Profile* should be developed regarding the current security level of the associated DNS components. Consider each of the categories and subcategories as it applies to your current DNS management and security policies and procedures. As with the cybersecurity framework itself, you may have additional processes or desired outcomes for consideration in your implementation.

Risk Assessment The next major step in the process comprises the risk assessment for affected DNS components. This step entails enumeration of each possible threat event. A threat event is an event that upon occurrence could impact the network and business detrimentally. Threat events may include events beyond security-related threats such as natural or man-made disasters so you may wish to consider all possible threats to secure your network and DNS in particular.

For each identified threat, consider the likelihood of the threat event occurring as well as the impact on your network and business should the threat event occur. The likelihood of a given threat event may be estimated by considering known vulnerabilities that may be exploited by an attacker to instigate the threat event. It's useful to plot each threat on a graph where the *x*-axis relates the relative impact of the threat while the *y*-axis reflects its relative likelihood. The relative impact could be estimated in terms of resource unavailability or downtime, end user or customer dissatisfaction, and/or lost revenue. Plotting risks in this manner can help you prioritize for which risks more urgent remediation is required.

As you may conclude from Figure 1.3, Risk #4 (R4) has a relatively high likelihood and high impact. This risk should likely be mitigated with the highest priority. Risk #2 of slightly lower impact and less likelihood should be next. Even though Risk #1 has a higher likelihood than Risk #2, its impact is substantially less. By applying controls, the goal is to shift each unmitigated risk down and to the left to render a lower overall residual risk to the organization.

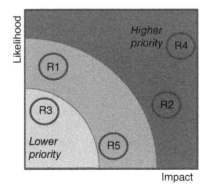

Figure 1.3. Risk Likelihood-Impact Plot

To add some structure to the process of assessing risk on a per system level, NIST published Federal Information Processing Standards (FIPS) Publication 199 (6). This publication defines standards for categorizing information and information systems based on the potential impact on the organization should certain threat events occur for use in assessing risk to an organization. Categorization is performed based on three security objectives for information and information systems.

- Confidentiality – the protection of information from unauthorized disclosure
- Integrity – protection against unauthorized modification or destruction of information
- Availability – protection against disruption of access to or use of information or system

FIPS Publication 199 defines three levels of impact on an organization for each of these objectives as follows:

- Low impact – expected to have a limited effect on the organization's operations, assets, or individuals; for example, loss of confidentiality, integrity, or availability could degrade an organization's capability though with noticeably reduced effectiveness. It could also result in minor damage to some or all of the organization's assets, minor financial loss and/or minor harm to individuals.
- Moderate impact – expected to have serious adverse impact on the organization's operations, assets, or individuals; for example, loss of confidentiality, integrity, or availability could cause significant degradation of an organization's capability though with substantially reduced effectiveness. It could also result in significant damage to the organization's assets, significant financial loss, and/or significant but not life-threatening harm to individuals.
- High impact – expected to have severe or catastrophic impact on the organization's operations, assets, or individuals; for example, loss of confidentiality, integrity, or availability could cause severe degradation of an organization's capability including the inability to perform one or more of its primary functions. It could also result in major damage to the organization's assets, major financial loss, and/or severe or catastrophic and life-threatening harm to individuals.

Categorization of each of the three objectives as low, moderate, or high is performed on various types of information at rest (e.g., within a file on a server) or in motion (e.g., within an IP packet traversing your network) through a network and on information systems themselves (e.g., servers, laptops, etc.). Examples of information types might be published DNS zone information or DNS query transaction information. The security categorization (SC) for the types and systems within your organization is represented as a tuple as illustrated in the following example:

$$SC_{\text{info type/system}} = \{(\textbf{confidentiality}, \text{LOW}), (\textbf{integrity}, \text{HIGH}),$$
$$(\textbf{availability}, \text{MODERATE})\}$$

For DNS, generally the highest requirement for most organizations is integrity, protecting DNS data from unauthorized changes. After all, users are relying on DNS data to enable them to connect to an intended destination. High availability likewise is paramount so that the resolution process and data is available. Confidentiality is typically lower in relative priority since DNS data generally is public information. However, many organizations publish a set of DNS information for resolution only for all or certain internal users and prohibit access for external users. In this scenario, for this type of information, confidentiality might be considered moderate if not high.

Target Profile and Security Planning   Your risk assessment will provide valuable input when prioritizing security initiatives in your security plan. We've included a sample of a DNS-specific framework core in Appendix A as a starting point. You can use our example framework core or create your own. Creating a target profile using the framework core allows you to define the desired outcomes for each of the defined categories along with those you may choose to add in. The differences between your target profile and your current profile define your to-do list of tasks, implementations, and process improvements necessary to transition from your current security implementation state to your desired target state.

To mitigate a given risk, a control or set of controls may be implemented to minimize the likelihood and/or impact of a given risk. Your risk assessment results will enable you to prioritize resources and efforts to apply controls in order to mitigate higher impact and higher likelihood threat events. A *control* is an implementation of technology, processes, and/or people resources that is intended to reduce such risks. Generally, a residual risk remains, which if excessive, may behoove you to apply additional controls.

In general, the application of multiple controls yields a *defense in depth* security approach that provides multiple lines of defense for a given threat event. Should an attacker penetrate one control, another is provided to hamper the further progress of the attack. When considering a given host, for example, a DNS server, a defense in depth approach entails securing each of the layers defined in Table 1.2.

Several aspects of this defense in depth strategy are common across several elements of your network, for example, all servers require strong credentials and all remote administrator access must be encrypted. Such *common controls* provide consistent protection and should apply to your DNS servers as well.

DNS-specific controls such as those example attributes outlined above provide added protection. The NIST framework core implicitly recommends a defense in depth strategy. NIST has also published a DNS-specific guide for secure DNS deployment (7). This useful guide describes DNS-specific controls with a particular focus on securing the integrity of DNS data. This guide provides thorough procedures for securing a BIND DNS server, including configuration and management of DNS security extensions (DNSSEC). We'll refer to this guide as well throughout this book where appropriate.

TABLE 1.2   Defense in Depth Layers

Data at rest	Data residing on the host, e.g., a file on a hard drive, thumb drive, or database	Configuration files, zone files, resource records, cached data
Data in transit	Data sent or received by the host	DNS queries and responses, DNS updates, zone transfers, configuration updates
Application	Reputability of each application running on the host	ISC BIND, Unbound, NSD, PowerDNS, Knot DNS, etc., i.e., your deployed DNS server application(s)
Host hardware and operating system	Reputability of the hardware manufacturer, software (e.g., BIOS) manufacturer, kernel and operating system hardening tactics	DNS server hardware, kernel, and operating system
Internal network	Internal firewalls, host firewalls, malware presence within internal infrastructure	Permissible ports and protocols for DNS, DNS ACLs, and port ACLs
Network perimeter	Boundary between trusted and untrusted environments	Permissible ports and protocols for DNS traffic traversal
External network	Internet-based vulnerabilities	Inbound purported DNS traffic; external DNS hosting providers; domain registrar(s)
Physical security	Building/datacenter/computer access, access control, property removal policies	DNS server physical security
Operations	Adherence to security policies by people, processes, and technologies; policy verification and enforcement	DNS configuration and transaction audits; training, holistic security awareness

Your security plan should define specific control implementations designed to mitigate specific threat events. Because each planned implementation will require organizational resources with respect to personnel involvement and perhaps capital and/or expense, you will generally need to prioritize and/or implement the plan in stages over time as resources permit. Application of the NIST cybersecurity framework provides structure and common language for efficiently conveying security status and goals. It also facilitates prioritization of security gaps to enable staging of the implementation of DNS and network security controls.

WHAT'S NEXT

Chapters 2 and 3 provide an overview of how DNS works with details regarding the DNS protocol, respectively. Chapter 4 introduces major security-related threats to and vulnerabilities of the DNS system. Chapters 5–11 delve more deeply into each vulnerability and defines detection and mitigation strategies accordingly. Chapter 12 discusses an overall security management architecture in terms of monitoring and maintaining your security approach and in defining response policies to security incidents. Chapter 13 discusses particular uses of DNS within broader security initiatives such as anti-spam and certificate validation.

2

INTRODUCTION TO THE DOMAIN NAME SYSTEM (DNS)

DNS OVERVIEW – DOMAINS AND RESOLUTION

As we introduced in Chapter 1, DNS is a foundational element of IP communications. DNS provides the means for improved usability of IP applications, such as insulating end users from typing IP addresses directly into applications like web browsers and enabling web servers to serve web pages compromised of diverse linked content. To communicate over an IP network, an IP device needs to send IP packets to the intended destination; and each IP packet header requires both source and destination IP addresses. DNS provides the translation from a user-entered named destination, for example, web site www address, to its IP address such that the sending device may populate the destination IP address with the address corresponding to the entered domain name.

DNS is not only useful for Internet users but also for network administrators. By publishing name-to-IP address mappings in DNS, the administrator gains the freedom to change IP addresses as needed for network maintenance, service provider changes, or general renumbering without affecting how end users connect. I can map my www address to 192.0.2.55 today and change it to 198.51.100.23 tomorrow without affecting how users reach my website. Of course, I'd need to keep both addresses active for some time given the caching of the former mapping in various recursive servers that had queried for my address before I made the change.

DNS Security Management, First Edition. Michael Dooley and Timothy Rooney.
© 2017 by The Institute of Electrical and Electronic Engineers, Inc. Published 2017 by John Wiley & Sons, Inc.

This level of indirection is also useful when configuring Internet "things" such as home appliances, smart cars, surveillance cameras, etc. that can connect to the Internet independently of user-initiated commands. Such connections may be initiated first by a DNS lookup of its configured web address, to which it connects to communicate a status update, alert, or other information. The thing's home website IP address may be changed as needed by changing its DNS entry in a manner similar to that just described.

As a network service, DNS has evolved from this simple domain name-to-IP address lookup utility to enabling very sophisticated "lookup" applications supporting voice, data, multimedia, and even security applications as we shall discuss in Chapter 13. DNS has proven extremely scalable and reliable for such lookup functions. We'll discuss how this lookup process works after first introducing how this information is organized.

Domain Hierarchy

The global domain name system is effectively a distributed hierarchical database. Each "dot" in a domain name indicates a boundary between tiers in the hierarchy, with each name in between dots denoted as a *label*. The top of the hierarchy, the "." or *root* domain provides references to top-level domains, such as .com, .net, .us, .uk, which in turn reference respective subdomains. Each of these top-level domains or TLDs is a child domain of the root domain. Each TLD has several children domains as well, such as ipamworldwide.com with the ipamworldwide domain beneath the com domain. And these children may have child domains and so on.

As we read between the dots from right to left, we can identify a unique path to the entity we are seeking. The text left of the leftmost dot is generally* the host name, which is located within the domain indicated by the rest of the domain name. A *fully qualified domain name* (FQDN) refers to this unique full [absolute] path name to the node or host within the global DNS data hierarchy. Figure 2.1 illustrates a FQDN mapping to the tree-like structure of the DNS database. Note that the trailing dot after .com. explicitly denotes the root domain within the domain name, rendering it fully qualified. Keep in mind that without this explicit FQDN trailing dot notation, a given domain name may be ambiguously interpreted as either fully qualified or relative to the "current" domain. This is certainly a legal and easier shorthand notation, but just be aware of the potential ambiguity.

NAME RESOLUTION

To illustrate how domain information is organized and how a DNS server leverages this hierarchical data structure, let's take a look at an example name resolution. Let's

---

* Some environments allow dots within hostnames which is relatively uncommon though permissible.

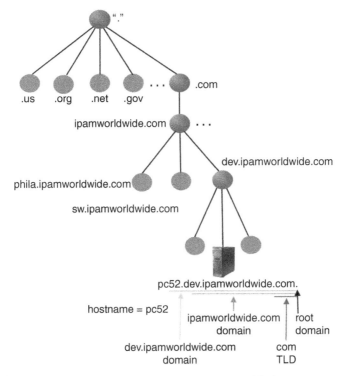

Figure 2.1. Domain Tree Mapping to a Fully Qualified Domain Name

assume you'd like to connect to a website, www.example.com. You'd enter this as your intended destination in my browser. The application (browser in this case) utilizes the sockets* application programming interface (API) to communicate with a portion of code within the TCP/IP stack called a *resolver*. The resolver's job in this instance is to translate the destination domain name you entered into an IP address that may be used to initiate IP communications. If the resolver is equipped with a cache and the answer resides in cache, the resolver responds to the application with the cached data and the process ends.

If not cached, the resolver issues a query for this domain name to the local recursive DNS server, requesting the server provide an answer. The IP address of this local DNS server is configured either manually or via DHCP using the domain server's option (option 6 in DHCP and option 23 in DHCPv6). This DNS server will then attempt to answer the query by looking in the following areas in the specified order and as illustrated in Figure 2.2. We recommend you specify at least two local DNS

---

* This API call is from the application to the TCP/IP layer of the protocol stack. The gethostbyname sockets/Winsock call initiates this particular process.

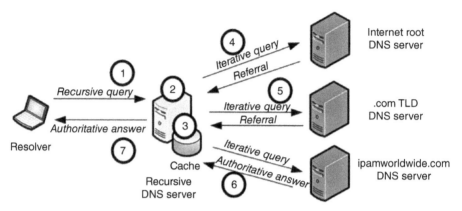

Figure 2.2.   Recursive and Iterative Queries in Name Resolution

server addresses to provide resilient resolution services in the event of a single DNS server failure.

We refer to this DNS server to which the resolver issues its query as a *recursive server*. "Recursive" means that the resolver would like the DNS server to try to find the answer to its query even if it does not know itself. From the resolver's viewpoint, it issues one query and expects an answer. From the recursive DNS server's perspective, it attempts to locate the answer for the resolver. The recursive server is the resolver's "portal" into the global domain name system. The recursive server accepts recursive queries directly from client resolvers and performs the following steps to obtain the answer to the query on behalf of the resolver.

1. The resolver initiates a query to resolve www.example.com to the recursive DNS server. As mentioned, the resolver knows which DNS server to query based on configuration via manual entry or via DHCP.

2. The queried server will first search its configured data files. That is, the DNS server is typically configured with resource record information for which it is *authoritative*. This information is typically configured using text files, a Microsoft Windows interface, or an IP address management (IPAM) system. For example, your company's DNS servers are likely configured with resolution information for your company's IP devices. As such, this is authoritative information. If the answer is found, it is returned to the resolver and the process stops.

3. If the queried server is not authoritative for the queried domain and is configured to enable recursion, it will access its cache to determine if it recently received a response for the same or similar query from another DNS server during a prior resolution task. If the answer for www.example.com resides in

cache*, the DNS server will respond to the resolver with this non-authoritative information and the process stops. The fact that this is not an authoritative answer is generally of little consequence, but the server alerts the resolver to this fact in its response.

4. If the queried DNS server cannot locate the queried information in cache, it will then attempt to locate the information via another DNS server that has the information. There are three methods used to perform this "escalation."

   a. If the queried recursive server is configured to forward queries to another server, the server will forward the query as configured in its configuration or zone file and will await a response.

   b. If forwarding is not configured and the cache information referenced in step 3 indicates a partial answer to the query, it will attempt to contact the source of that information to locate the ultimate source and answer. For example, a prior query to another DNS server, server X, may have indicated that DNS server X is authoritative for the `example.com` domain. The recursive DNS server may then query DNS server X for information leading to resolution of `www.example.com`.

   c. If no information is found in cache, the server cannot identify a referral server, or forwarding did not provide a response[†] or is not configured, the DNS server will access its *hints* file. The hints file provides a list of *root name servers* to query in order to begin traversing down the domain hierarchy to a DNS server that can provide an answer to the query.

   Upon querying either a root server or a server further down the tree based on cached information, the queried server will either resolve the query by providing the requested IP address(es) for `www.example.com`. or will provide a referral to another DNS server further down the hierarchy "closer" to the sought-after FQDN. For example, upon querying a root server, you are guaranteed that you will not obtain a direct resolution answer for `www.example.com`. However, the root name server will *refer* the querying DNS server to the name servers that are authoritative for `com` as illustrated in Figure 2.2. The root servers are "delegation-only" servers and do not directly resolve queries, only answering with delegated name server information for the queried TLD.

5. The recursive server *iterates*[‡] additional queries based on responses down the domain tree until the query can be answered. Continuing with our example,

---

* Cache entries are temporary and are removed by DNS servers based on user configuration settings as well as advertised lifetime ("time to live," TTL) of a resource record.
[†] If the `forward only` option is configured, the resolution attempt will cease if the forwarded query returns no results; if the `forward first` option is configured, the process outlined in this paragraph ensues, with escalation to a root server.
[‡] These "point-to-point" queries are referred to as iterative queries.

the recursive server queries one of the .com name servers referred to by the root server. The answer received from that name server will also be a referral to the name server that is authoritative for example.com., and so on down the tree.

Note that by issuing queries to other DNS servers to locate resolution information, the recursive server itself performs a resolver function to execute this lookup. The term *stub resolver* is commonly used to identify resolvers which do not iterate down the domain tree. Stub resolvers such as those within typical end user clients are configured only with which recursive name servers to query.

6. Upon receipt of the referral from the TLD server indicating two or more name servers and corresponding IP addresses that are delegated to example.com, the recursive server prepares its next iterative query to issue to one of these servers. The response includes the answer to the query in the form of one or more resource records matching the queried name, class, and resource record type.

   The recursive server generally updates its cache not only with the ultimate answer for the specific query, but with any additional information provided with the answer and referral messages received in the process. In this way, the recursive server caches the name server domain names and IP addresses for the .com and example.com domains. When the same or another stub resolver queries for another domain with the .com domain subtree, the recursive server can utilize its cache to query one of the .com name servers directly without needing to query the root server as mentioned in step 4 previously. If an answer cannot be found, the recursive server will also cache this "negative" information as well for use in responding to similar queries.

7. When the answer is received, the recursive DNS server will provide the answer to the stub resolver as per Figure 2.2 which can update its cache and the process ends. It might seem that this process is laborious and time consuming, but as you've probably experienced, it all happens very quickly. Root and TLD servers are "delegation-only" servers in that they only support referrals. This helps performance, as does caching by the recursive DNS server, which can leverage cached DNS information to streamline the resolution process.

   As mentioned earlier, the stub resolver may be configured to cache this information as well. In this case, only the answer is cached, not other domain tree nodes, because the stub resolver only queries its configured name servers. But for users frequently visiting common sites, the stub resolver cache can improve application response time.

We mentioned the use of DNS forwarders in step 4, where the server queried by stub resolvers forward all or some queries to other DNS servers for recursion. You might deploy this configuration if you have a consolidated set of internal caching servers through which you funnel external-bound DNS queries or if you're using a third-party DNS resolver service. In both cases, you can configure your internal forwarding DNS servers to resolve internal name space and forward other queries to your caching servers or to the service providers' DNS servers, respectively. The forwarding DNS server generally caches resolution answers like recursive servers to minimize DNS message flow and improve resolution performance.

In summary, the resolution process entails (a) finding a set of name servers with authoritative information to resolve the query in question, and (b) querying an authoritative server for the desired information. In our example, the desired information is the IP address corresponding to the host domain name `www.example.com`. This "translation" information, mapping the queried domain name to an IP address, is stored in the DNS server in the form of a resource record. Different types of resource records are defined for different types of lookups. Each resource record contains a "key" or lookup value and a corresponding resolution or answer value. In some cases, a given lookup value for a given type may have multiple entries in the DNS server configuration. In this case, the authoritative DNS server will respond with the entire set of resource records, or *resource record set* (*RRSet*), matching the queried value (name), class, and type.

The organization of DNS data is such that DNS servers are configured at all levels of the domain tree as authoritative for their respective domain information, including where to direct queriers further down the domain tree. In most cases, these servers at different levels are administered by different organizations. Not every level or node in the domain tree requires a different set of DNS servers as an organization may serve multiple domain levels within a common set of DNS servers.

While the top three layers of the domain tree typically utilize three sets of DNS servers under differing administrative authority, the support of multiple levels or domains on a single set of DNS servers is a deployment decision. This decision hinges primarily on whether administrative delegation is required or desired. For example, the DNS administrators for the ipamworldwide.com domain may desire to retain administrative control of the sw.ipamworldwide.com. domain, but to delegate dev.ipamworldwide.com to a different set of administrators and name servers. This leads us to a discussion regarding the distinction between zones and domains.

ZONES AND DOMAINS

The term *zone* is used to differentiate the level of administrative control with respect to the domain hierarchy. In our example, the ipamworldwide.com zone contains authority for the ipamworldwide.com, sw.ipamworldwide.com, and phila.ipamworldwide.com domains, while the dev.ipamworldwide.com zone retains authority for the dev.ipamworldwide.com domain and its subdomains as illustrated in Figure 2.3.

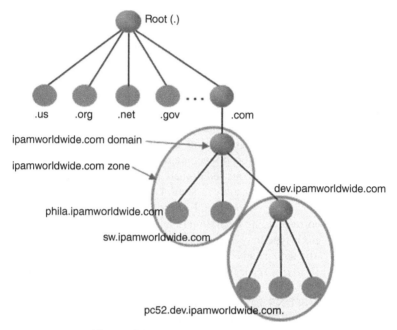

Figure 2.3. Zones as Delegated Domains

By delegating authority for dev.ipamworldwide.com, the DNS administrators for ipamworldwide.com agree to pass all resolutions for dev.ipamworldwide.com (and below in the domain tree for any subdomains of dev.ipamworldwide.com) to DNS servers administered by personnel operating the dev.ipamworldwide.com zone. These dev.ipamworldwide.com administrators can manage their domain and resource records and any children autonomously; they just need to inform the parent domain administrators (for ipamworldwide.com) where to direct queries they receive as resolvers or other DNS servers attempt to traverse down the domain tree seeking resolutions.

Thus, administrators for the ipamworldwide.com zone must configure all resource records and configuration attributes for the ipamworldwide.com zone, including subdomains within the ipamworldwide.com zone such as the sw.ipamworldwide.com domain. At the same time, ipamworldwide.com administrators must provide a delegation linkage to any child zones, such as dev.ipamworldwide.com. This delegation linkage is supported by entering name server (NS) resource records within the ipamworldwide.com zone file, which indicate which name servers are authoritative for the dev.ipamworldwide.com delegated zone. These NS records provide the continuity to delegated child zones by referring resolvers or other name servers further down the domain tree. Corresponding A or AAAA records called *glue records* are also typically defined to glue the resolved NS host domain name to an IP address to enable direct addressing of further queries.

The encircled domain nodes in Figure 2.3 indicate that the ipamworldwide.com zone contains the ipamworldwide.com domain plus two of its children, phila and sw. The ipamworldwide.com DNS administrators are responsible for maintaining all DNS configuration information for the ipamworldwide.com zone including subdomains managed within the ipamworldwide.com zone, as well as referrals to DNS servers serving delegated child zones, such as dev.ipamworldwide.com as shown in Figure 2.3. Thus, when other DNS servers around the world are attempting to resolve any domain names ending in ipamworldwide.com on behalf of their clients, their queries will require traversal of the ipamworldwide.com DNS servers and perhaps other DNS servers, such as those serving the dev.ipamworldwide.com zone.

The process of delegation of the name space enables autonomy of DNS configuration while providing linkages via NS record referrals within the global DNS database. As you can imagine, if the name servers referenced by these NS records are unavailable, the domain tree will be broken at that point, inhibiting resolution of names at that point or below in the domain tree. If the dev.ipamworldwide.com DNS servers are down, authoritative resolution for dev.ipamworldwide.com *and its children* will fail. This illustrates the requirement that each zone must have at least two authoritative DNS servers for redundancy.

In summary, the administrators for the ipamworldwide.com zone will configure their DNS servers with configuration and resource record information for the ipamworldwide.com, phila.ipamworldwide.com, and sw.ipamworldwide.com domains. They will also need to configure their servers with just the names and addresses of DNS servers serving delegated child zones, particularly dev.ipamworldwide.com in this case. They need know nothing further about these delegated domains; just where to refer the querying recursive DNS server during the resolution process.

Dissemination of Zone Information

Given the criticality of the DNS service in resolving authoritatively and maintaining domain tree linkages, DNS server redundancy is a must. DNS server configuration information consists of server configuration parameters and declarations of all zones for which the server is authoritative. This information can be defined on each server that is authoritative for a given set of zones. Additions, changes, and deletions of resource records, the discrete resolution information within each zone configuration file, can be entered once on a *master* server, or more correctly, the server that is configured as master for the respective zone. The other servers that are likewise authoritative for this information can be configured as *slaves* or *secondaries*, and they obtain zone updates by the process of *zone transfers*. Zone transfers enable a slave server to obtain the latest copy of its authoritative zone information from the master server. Microsoft Active Directory-integrated DNS servers support zone transfers for compatibility with this standard process, but also enable DNS data replication using native Active Directory replication processes.

Versions of zone files are tracked by a zone serial number which must be updated every time a change is applied to the zone. Slaves are configured to periodically check the zone serial number set on the master server; if the serial number is larger than its own value defined for the zone, it will conclude that it has outdated information and will initiate a zone transfer. Additionally, the server that is master for the zone can be configured to *notify* its slaves that a change has been made, stimulating the slaves to immediately check the serial number and perform a zone transfer to obtain the updates more quickly than awaiting the normal periodic update check.

Zone transfers may consist of the entire zone configuration file, called an absolute zone transfer (AXFR) or of the incremental updates only, called an incremental zone transfer (IXFR). In cases where zone information is relatively static and updated from a single source, for example, an administrator, the serial number checking with AXFRs as needed works well. These so-called *static zones* are much simpler to administer than their counterpart: *dynamic zones*. Dynamic zones, as the name implies, accept dynamic updates, for example, from DHCP servers updating DNS with newly assigned IP addresses and corresponding host domain names. Updates for dynamic zones can utilize IXFR mechanisms to maintain synchronization among the master and multiple slave servers.

The popular BIND DNS reference implementation utilizes journal files on each server to provide an efficient means to track dynamic updates to zone information. These journal files are temporary appendages to corresponding zone files and enables tracking of dynamic updates until the server writes these journal entries into the zone file and reloads the zone. Many server implementations load the zone file information into memory along with incremental zone updates, which are also loaded into memory for fast resolution. Other approaches to storing zone information include use of a database such as is the case with the PowerDNS and Knot DNS implementations.

Additional Zones

Root Hints   We mentioned a *hints file* during the overview of the resolution process. This file should provide a list of DNS server names and addresses (in the form of NS, A, and AAAA resource records) that the server should query if the resolver query cannot be resolved via authoritative, forwarded, or cached data. The hints file will typically list the Internet root servers, which are authoritative for the Internet root (.) of the domain tree. Querying a root server enables the recursive server to start at the top to begin the traversal down the domain tree in order to locate an authoritative server to resolve the query. The contents of the hints file for Internet root servers may be obtained from www.internic.net/zones/named.root, though all major DNS server implementations include this file with their distributions.

Some environments may require use of an internal set of root servers, where Internet access is restricted by organizational policy. In such cases, an internal version of the hints file can be used, listing names and addresses of internal root servers instead of the Internet root servers as we'll discuss in Chapter 5. The organization itself would need to maintain the listing of internal root servers, as well as their requisite root zone configurations.

Localhost Zones   Another zone file that proves essential is the localhost zone. The localhost zone enables one to resolve "localhost" as a hostname on the given server. A corresponding in-addr.arpa. zone file resolves the 127.0.0.1 loop-back address. A single entry within the 0.0.127.in-addr.arpa zone maps this loopback address to the host itself. This zone is required as there is no upstream authority for the 127.in-addr.arpa domain or subdomains. Likewise, the IPv6 equivalents need to be defined for the corresponding IPv6 loopback address,::1. The localhost zone simply maps the localhost hostname to its 127.0.0.1 or::1 IP address using an A and AAAA record, respectively. Please refer to *IP Address Management Principles and Practice* (8) for a full discussion of this and reverse zones in general.

RESOLVER CONFIGURATION

DNS does require some basic client configuration prior to use. This initial configuration may be performed manually or by obtaining this information from a DHCP server. The client configuration consists of configuring resolver software regarding which DNS server(s) to query for resolution.

Figure 2.4 illustrates the configuration of a Microsoft Windows resolver in terms of manually defining the DNS server to query or the use of DHCP to obtain DNS server addresses automatically. Though your screen shots may look a bit different, these screens for Windows have looked nearly identically over the last several versions of Windows.

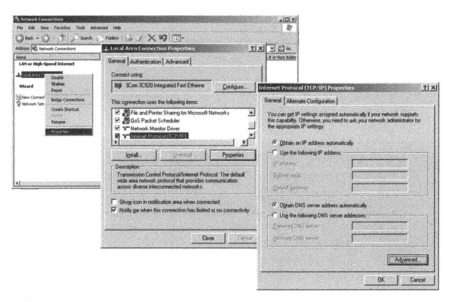

Figure 2.4.   Microsoft Windows Configuration of IP Address DNS Servers to Query

Microsoft Windows enables entry of multiple DNS servers to query within its graphical interface. Notice there are two entries in the "brute force" method shown on the screen on the right of Figure 2.4, one for preferred and another for alternate. Clicking the Advanced tab enables entry of more than two and in particular order. We recommend having *at least* two DNS servers configured for the resolver in the event one DNS server is unavailable, the resolver will automatically query the alternative server. If the "Obtain DNS server address automatically" radio button is selected, as shown in Figure 2.4, the resolver will obtain a list of DNS servers typically via DHCP.

On Unix or Linux based systems, the /etc/resolv.conf file can be edited to configure the resolver. The key parameter in this file is one or more nameserver statements pointing to DNS servers, but a number of options and additional directives enable further configuration refinement as described below. The italicized text should be replaced by actual data referenced; for example, *domain* should be replaced with a DNS domain name.

- nameserver *IP_address* – the IP address of a recursive DNS server to query for name resolution; multiple nameserver entries are allowed and encouraged. The nameserver entry instructs the resolver where to direct DNS queries.

- domain *domain* – the DNS domain where this host (on which this resolver is installed) resides. This is used when resolving relative hostnames, as opposed to fully qualified host domain names.

- search *domain(s)* – the search list of up to six domains in which to search the entered hostname for resolution. Thus if we type in www for resolution, the resolver will successively append domains configured in this parameter in an attempt to resolve the query. If the entry search ipamworldwide.com. exists in resolv.conf, entry of www will result in a resolution attempt for www.ipamworldwide.com.

- sortlist *address/mask* list – enables sorting of resolved IP addresses in accordance with the specified list of address/mask combinations. This enables the resolver to choose a "closer" destination if multiple IP addresses are returned for a query.

- options – keyword preceding the following which enables specification of corresponding resolver parameters including the following:

  o debug – turns on debugging

  o ndots *n* – defines a threshold for the number of dots within the entered name required before the resolver will consider the entered name simply a hostname or a qualified domain name. When considered a hostname, the hostname will be queried as appended with domain names specified within the domain or search parameter.

  o timeout *n* – number of seconds to wait before timing out a query to a DNS server.

- attempts *n* – number of query attempts before considering the query a failure.
- rotate – enables round robin querying among DNS servers configured within the nameserver directives. Queries will be sent to a different server each time and cycled through.
- no-check-names – turns off name checking of entered host names for resolution. Normally, underscore characters are not permitted, for example, so setting this option enables query processing to proceed without validation of the entered hostname.
- inet6 – causes the resolver to issue a query for a AAAA record to resolve the entered hostname before attempting an A record query.

search and options settings can also be overridden on a per process basis via corresponding environment variable settings.

SUMMARY

This chapter described the organization of DNS data in a tree structure where each node of the tree may be managed by independent entities. We also discussed the various formats for looking up information as well as the basic types of zone files generally configured on DNS servers. Lastly we provided a brief overview of device resolver configuration. This background serves as a foundation for understanding the structure of "data at rest" on DNS servers and resolvers, as well as the process for locating DNS information within the global DNS tree. Now onto the next chapter to review DNS data in motion, the DNS protocol.

3

DNS PROTOCOL AND MESSAGES

Attackers may leverage the DNS protocol itself to launch attacks or attempt to infiltrate DNS data in motion. To help us understand such techniques, this chapter provides an overview of DNS protocol operations and message formats.

DNS MESSAGE FORMAT

Encoding of Domain Names

In Chapter 2, we discussed the organization of DNS information into a domain hierarchy as well as the basics of how a client or resolver performs resolution by issuing a recursive query to a DNS server which in turn iterates the query in accordance with the domain hierarchy to obtain the answer to the query. Next we'll dig deeper into the DNS query and general message format, but we'll first introduce the representation of domain names within DNS messages. Domain names are formatted as a series of *labels*. Labels consist of a one byte length field followed by that number of bytes/ASCII characters representing the label itself. This sequence of labels is terminated by a length field of zero indicating the root "." domain. For example, the series of labels for www.ipamworldwide.com would look like the following in ASCII format. Length bytes are highlighted in darker shading in Figure 3.1.

DNS Security Management, First Edition. Michael Dooley and Timothy Rooney.
© 2017 by The Institute of Electrical and Electronic Engineers, Inc. Published 2017 by John Wiley & Sons, Inc.

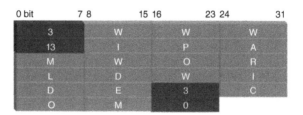

Figure 3.1.   DNS Labels

Starting at the upper left, the value "3" of the first length byte indicates that the following three bytes comprise first label, "www." The fifth or next byte after this is our next length byte, which has a value of "13" (0xD*), which is the length of "ipamworldwide." After this label, the following byte of value "3" is the length of "com." Finally, the zero-value byte indicates the root "." domain, fully qualifying the domain name. Note that the darker shaded bytes in the figure are encoded as length bytes to differentiate them from host or domain name characters containing numbers. The first byte in a name will almost always[†] be a length byte followed by that number of bytes representing the first label, immediately followed by another length byte to eliminate ambiguity.

Name Compression

A given DNS message may contain multiple domain names, and many of these may have repetitive information, for example, the ipamworldwide.com. suffix. The DNS specification enables message compression in order to reduce repetitive information and thereby reduce the size of the DNS message. This works by using *pointers* to other locations within the DNS message that specify a common domain suffix. This domain suffix is then appended at the point of location referenced by the pointer.

Let's say, for example, that our query for www.ipamworldwide.com. returns a pair of DNS servers that can be queried for more information: ns1.ipamworldwide.com. and ns2.isp.com. The ipamworldwide.com. portion of these domain names is common to the query and one of the answers, while only the .com portion is common to the question, the first answer, and the second answer.

Thus, the message is formulated by fully specifying the domain name www.ipamworldwide.com. as illustrated above in Figure 3.1. Then, when specifying ns1, instead of fully specifying ns1.ipamworldwide.com, only ns1 is specified, followed by a pointer to the ipamworldwide.com. suffix earlier in

---

* The "0x" prefix indicates the string following is in hexadecimal format; '0b' indicates binary format.
† As we'll discuss next, the length byte may alternatively consist of a two-byte pointer or a DNS extension label.

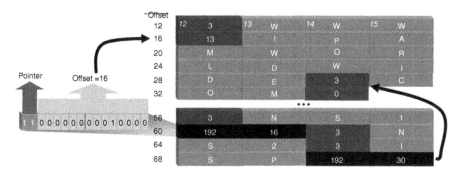

Figure 3.2. Name Compression with Pointers

the message. When identifying `ns2.isp.com`, the `ns2.isp` labels are specified, followed by a pointer to the `.com` suffix within the message.

How do DNS resolvers and servers differentiate a pointer from a standard label length byte? The DNS standard stipulates that each label may be of length 0–63 bytes. In binary, this is 000000000–00111111. Thus, the first two bits, 0b00 in this case, identify the byte as a standard-length byte, indicating the length of the following label. A pointer is identified by setting the first two bits to 0b11, and is comprised of two bytes, where the 0b11 bits are followed by 14 bits identifying the offset in bytes from the beginning of the DNS header. The first byte of the DNS message header is considered byte 0, and as the message is created, pointers are defined pointing to byte offsets from this point.

Let's look at how this maps out from our prior example. Let's say that beginning 12 bytes from the DNS header, we've included the domain name, `www.ipamworldwide.com`. Now, later in the message, beginning at byte 56 from the beginning of the header, we would like to encode responses `ns1.ipamworldwide.com` and `ns2.isp.com`.

Figure 3.2 indicates how this would look. The first portion is as we discussed earlier, with length bytes (dark shading) followed by the respective number of name bytes (light shading). At byte position 56 in our example, the `ns1` portion of the name is encoded normally, using a label length of "3," followed by `ns1`. However, the next byte is not a standard-length byte but a pointer "double-byte" as it begins with 0b11 and is shown as shaded black in the figure. The value encoded in the 14-bit offset field of the pointer is "16," indicating that the portion of the domain name starting at an offset of 16 bytes from the start of the DNS header should be appended to the `ns1` label already specified. The first row of bytes in the figure below enumerates the individual byte offsets (italics), and byte 16 is the length byte of value "13," followed by encoding for `ipamworldwide`, followed by a length byte of value "3," then `com`, then `.` (length byte of value "0"). Concatenating this together, we arrive at the result: `ns1.ipamworldwide.com`.

Returning to the next domain name after processing the pointer, we find encoding for ns2.isp followed by a pointer to byte offset 30, which points to the length byte of "3," followed by com., completing the domain name as ns2.isp.com. Considering just these three example domain names, the number of bytes in the message occupied by these domain names can be compressed from 59 bytes to 39 bytes.

Internationalized Domain Names

DNS resolvers and servers communicate hostname queries and responses in ASCII-formatted messages. Configuration information is stored in ASCII text files. Unfortunately, while ASCII characters have been defined to effectively represent the English language, they do not enable formatting of characters from other languages, especially those using a non-Latin-based alphabet. This limitation certainly impacts the ease-of-use of IP applications in countries where people do not use the English language. RFC 5890 (9) is a standards track RFC that addresses this limitation.

The RFC is entitled Internationalizing Domain Names in Applications (IDNA). The "in applications" qualifier in the title insinuates the involvement of applications in this process. Indeed, the onus is placed on the application, such as a web browser or email client, to convert the user's native language entry into an ASCII-based string that can be communicated to a DNS server for resolution. This ingenious approach enables application level support of international character sets for end users without affecting the DNS protocol (or other ASCII-based IP protocols like SMTP either). Existing DNS servers can be configured to resolve these ASCII-encoded domain names as they would for native ASCII-based domain names.

International character sets are encoded as Unicode characters. The Unicode standard "provides a unique number for every character, no matter what the platform, no matter what the program, no matter what the language," according to the Unicode Consortium website (www.unicode.org). Every character is represented as a unique 2–3 byte hexadecimal number. RFC 5890, and its related RFCs 5891 (10), 5892 (11), 5893 (12), 5894 (13), and 5895 (14), describe the process of converting a Unicode-based domain name to an ASCII-formatted domain name. Note that technically the domain labels are each converted, not the "domain name." Recall that a domain name is communicated using a series of domain labels, one for each domain level or text string between the dots, each encoded with a length byte beginning with 0b00.

To resolve international domain names, a DNS server must be configured with resource records encoded in ASCII format, specifically Unicode-mapped ASCII characters referred to as *punycode*. The output of the punycode algorithm results in an ASCII string, which is then prefixed with the ASCII Compatible Encoding (ACE) header, xn–. Thus, within the DNS infrastructure, domains denoted as xn–<additional ASCII characters> are likely punycode representations of an internationalized domain name. The application, for example, web browser, is responsible

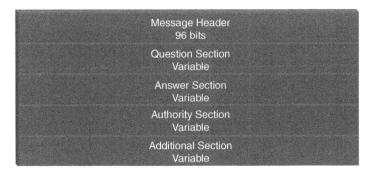

Figure 3.3. DNS Message Fields

for converting the user-entered URL into Unicode format, then into punycode. The punycode domain name is passed to the resolver on the client for resolution via DNS using ASCII characters. The punycode algorithm is specified in RFC 3492 (15) and several websites are available for performing conversions for entry into DNS.

Consider an example (16): Let's consider a web server host address in the żdżbło.com domain as www.żdżbło.com. The domain name contains diacritics and has characters outside of the ASCII character set. The web browser in which this URL is entered would convert this to ASCII characters or punycode as www.xn–dbo-iwa1zb.com. A corresponding A or AAAA record entry in DNS for the www.xn–dbo-iwa1zb.com. host would enable the end user to enter a native language URL while utilizing the existing base of DNS servers deployed throughout the world to identify and connect via the IP address of the destination web server. The net result is that these DNS messages sent on the wire are encoded as ASCII characters.

DNS Message Format

Now let's look more closely at the format of DNS messages used to perform this overall resolution function, incorporating the label-formatted domain names we discussed earlier. DNS messages are transmitted over UDP by default, using the well-known port 53. TCP can also be used on port 53. The basic format of a DNS message is illustrated below.

- The message header contains fields that define the type of message and associated information, including the number of records for each of the following fields.
- The Question section specifies the information being sought via this message.
- The Answer section contains zero or more resource records that answer the query specified in the Question section.

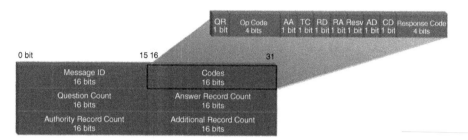

Figure 3.4.   DNS Message Header

- The Authority section contains zero or more resource records referring to name servers authoritative for the given answer or pointing to delegated name servers down the domain tree to which a successive iterative query may be issued.
- The Additional section contains zero or more resource records that contain supplemental information related to the Question but are not strictly answers to the Question.

Message Header   The DNS message header is included on every DNS message and conveys the type of message and associated parameters as illustrated in Figure 3.4.

The message header is comprised of six 16-bit fields.

- Message ID – also referred to as transaction ID, an identifier assigned by the resolver and copied in replies from the DNS server to enable resolver correlation of responses with queries.
- Codes – message codes germane to this message. We'll examine these code fields next.
- Question Count (QDCOUNT) – the number of questions contained in the Question section of the DNS message.
- Answer Record Count (ANCOUNT) – the number of resource records contained in the Answer section of the DNS message.
- Authority Record Count (NSCOUNT) – the number of resource records contained in the Authority section of the DNS message.
- Additional Record Count (ARCOUNT) – the number of resource records contained in the Additional section of the DNS message.

The following Codes bits have been defined:

- QR – Query/Response – this flag indicates that this message is a query (0) or a response (1).

- Opcode – the operation code for this message. Presently, the following values have been defined:
  - 0 = Query
  - 1 = Reserved (formerly inverse query, now retired)
  - 2 = Server status request
  - 3 = Unassigned
  - 4 = Notify – enables a master zone server to inform a slave zone server with the same zone (and for a slave to acknowledge) that a change has been made to the zone data. For Notify messages, the Authority and Additional sections are not used and respective record counts in the DNS header should be set to 0.
  - 5 = Update – enables a DHCP server or other entity to update zone data on a DNS server. For Update messages, the interpretation of DNS message fields and corresponding header fields differs from that described above. The message format for Update messages is described in the DNS Update Messages section.
  - 6–15 = Unassigned
- AA – Authoritative Answer – when set by the responding server, this message contains an authoritative answer to the question. This means the response was derived from a DNS server that was configured with the zone's information. If it is not set, the answer was derived from a nonauthoritative DNS server, likely, cached information from a prior query. Where multiple answers are provided, this flag pertains to the first record in the Answer section. When set by the client on the query, this indicates that an authoritative answer (not cached) is required.
- TC – Truncated Response – this code set by the responding DNS server, indicates that this message was truncated for transmission. This is generally due to the packet length restrictions of UDP packets, the default transport layer protocol used by DNS. Responses received with the TC bit set typically stimulates the querier to reattempt the query over TCP.
- RD – Recursion Desired – this flag is set by the querying resolver to indicate that the querier would like the DNS server to iteratively resolve the query, traversing the domain tree as necessary. Most resolvers set this flag to indicate a query as a recursive query, while a DNS server will not usually set this flag when querying other servers, unless it is forwarding a query.
- RA – Recursion Available – this flag indicates that recursive query service is available from this responding DNS server.
- Reserved or Z bit – Reserved (0).
- AD – Authentic Data – used within the context of DNS security extensions (DNSSEC), this bit is set by a responding DNS server to indicate that

information within the Answer and Authority sections is authentic, meaning it has indeed been authenticated.

- CD – Checking Disabled – used within the context of DNSSEC, this bit enables a resolver to disable signature validation in a DNS server's processing of this particular query.
- RCODE – Response Code – provides result status to the client from the responding DNS server. The currently defined response codes are summarized in Table 3.1. Note that given the 4-bit RCODE field, decimal values 1–15 are encoded within the DNS header RCODE field.

The DNS extensions' (EDNS0, discussed later in this chapter) OPT resource record adds a capacity for eight additional RCODE bits, bringing the total to 12 bits (up to decimal value 4095) when used in combination with the header RCODE bits. TSIG and TKEY resource records define a 16-bit result ("error") field, which is mostly consistent with codes from the DNS header and the EDNS0 extended header (OPT record). The only inconsistency is in the interpretation of code 16 (0x10) which has similar interpretation depending on the context of its use within an OPT or TSIG record.

Question Section   Each of these questions is of the format illustrated in Figure 3.5. The Question section within the DNS message format contains, as you might have guessed, the question that is being asked for this query. This section can contain more than one question, as identified by the number referenced in the QDCOUNT header field. Each of these questions has the following format.

The QNAME field contains the domain name, formatted as a series of labels. The QTYPE field indicates the query type, or for what purpose is this question being asked. Any resource record type may be included, and we have summarized those currently defined in Appendix B. However, there are some QTYPE values that are unique to requesting zone transfers, for example, that are presently defined per Table 3.2.

The QCLASS field indicates for which class this query is being made; for example, IN for Internet class, the most common class. Classes essentially enable management of parallel namespaces. Currently defined QCLASSes (and DNS CLASSes) in general are defined in Table 3.3.

Answer Section   The Answer section contains zero or more answers in the form of resource records. The number of answers is specified in the ANCOUNT header field. We summarize the different types of resource records in Appendix B, and they all share a common generic format as defined below.

The **Name** field, also known as the Owner name field, is the lookup name corresponding to this resource record (corresponding to the lookup value or QNAME in the original question).

The **Type** field indicates the type of information that is provided for this name, for example, the RRType. For example, a type of "A" means that this resource record provides IPv4 address information.

TABLE 3.1 DNS Message Response Codes

RCODE				
Decimal	Hex	Name	Description	Reference
0	0	NoError	No errors	RFC 1035 (17)
1	1	FormErr	Format error – server unable to interpret the query	RFC 1035
2	2	ServFail	Server failure – server problem has prevented processing of this query	RFC 1035
3	3	NXDomain	Nonexistent domain – domain name does not exist	RFC 1035
4	4	NotImp	Not implemented – query type not supported by this server	RFC 1035
5	5	Refused	Query refused – server refused the requested query; e.g., refusal of a zone transfer request	RFC 1035
6	6	YXDomain	Name exists when it should not as determined during DNS update prerequisite processing	RFC 2136 (18), RFC 6672 (19)
7	7	YXRRSet	RR Set exists when it should not as determined during DNS update prerequisite processing	RFC 2136
8	8	NXRRSet	RR Set that should exist does not as determined during DNS update prerequisite processing	RFC 2136
9	9	NotAuth	Server is not authoritative for the zone listed in the Zone section of the DNS update message	RFC 2136
10	A	NotZone	Name used in the prerequisite or Update section of a DNS Update message is not contained in zone denoted by the Zone section of the message	RFC 2136

(continued)

TABLE 3.1 (Continued)

RCODE			
Decimal	Hex	Name	Description	Reference
11–15	B–F	Unassigned		
16	10	BADVERS	Unsupported (bad) OPT RR version	RFC 6891 (20)
16	10	BADSIG	TSIG signature failure	RFC 2845 (21)
17	11	BADKEY	Key not recognized	RFC 2845
18	12	BADTIME	Signature out of the valid server signature time window	RFC 2845
19	13	BADMODE	Invalid TKEY mode – requested mode not supported by this server	RFC 2930 (22)
20	14	BADNAME	Nonexistent or duplicate key name	RFC 2930
21	15	BADALG	Algorithm not supported	RFC 2930
22	16	BADTRUNC	Bad truncation – message authentication code (MAC) too short	RFC 4635 (23)
23	17	BADCOOKIE	Temporary – bad or missing server cookie	RFC 7873 (24)
24–3840	14–F00	Available for assignment		
3841–4095	F01–FFF	Reserved for private use		RFC 6895 (25)
4096–65534	1000–FFFE	Unassigned		
65535	FFFF	Reserved		RFC 6895

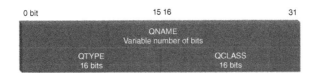

Figure 3.5.   Question Section Format

TABLE 3.2   DNS QTYPEs

QTypes Only	Query Purpose	QType ID (decimal)	IETF Status	Defining Document
*(ANY)	All resource records	255	Standard	RFC 1035
MAILA	Mail agent resource records	254	Experimental	RFC 1035
MAILB	Mailbox resource records	253	Obsolete	RFC 1035
AXFR	Absolute zone transfer (entire zone)	252	Standard	RFC 1035
IXFR	Incremental zone transfer (changes only)	251	Proposed standard	RFC 1995 (26)

TABLE 3.3   DNS Classes

Class				
Decimal	Hexadecimal	Name	Description	Reference
0	0	Reserved	Reserved	RFC 6895
1	1	IN	Internet	RFC 1035
2	2	Unassigned	N/A	IANA
3	3	CH	Chaos	RFC 1035
4	4	HS	Hesiod	RFC 1035
5–253	5–FD	Unassigned	N/A	IANA
254	FE	None	None	RFC 2136
255	FF	Any	Any class (valid as QCLASS but not on resource records)	RFC 1035
256–65279	100–FEFF	Unassigned	N/A	IANA
65280–65534	FF00–FFFE	Reserved for private use		RFC 6895
65535	FFFF	Reserved	Reserved	RFC 6895

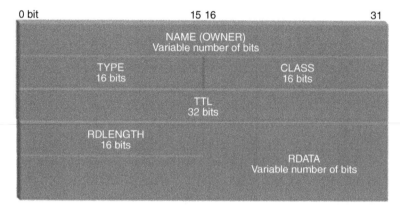

Figure 3.6.   Answer Section Format

The **Class** field represents the namespace class, such as IN for Internet. Valid classes are displayed in Table 3.3.

The **TTL** or Time-to-Live field provides a time value in seconds with respect to the valid lifetime of the resource record. The receiver of this information may cache this information for *TTL* seconds and may use it reliably. However, upon expiration of the TTL, the cached information must be discarded.

The **RDLength** field indicates the length in bytes of the results field, RData.

The **RData** field contains the corresponding information of the specified Type in the identified Class, for the given Owner. The RData field has a variable format based on the resource record type.

The Answer section essentially contains the resource records published within the corresponding zone file or database that match the question (Qname, Qclass, and Qtype). Resource records published within a zone file follow the same format just described, omitting the RDLength field which the server calculates and populates on the "wire" format of Figure 3.6.

Authority Section   The Authority section contains *NSCOUNT* number of answers in the form of resource records of the same format as discussed in the Answer section. Generally only NS (name server) resource records are valid within the Authority section, though most name servers return an SOA record in this section if the queried name server is authoritative but the Answer section is empty. This section also contains information about other name servers that are authoritative for the queried information. This information is used by the querying resolver or more likely, recursive name server, to determine the next name server to query in traversing the domain tree to find the ultimate answer.

Additional Section   The Additional section contains *ASCOUNT* number of answers in the form of resource records, which provide additional or related information to the query, in the same format as discussed in the Answer section.

DNS Update Messages

Update messages, also referred to as Dynamic DNS (DDNS) messages, enable a remote source, presumably an end user client, DHCP server, or other [hopefully] legitimate source to perform an update (add, modify, or delete) of one or more resource records within a zone. While Update messages utilize the same basic format as DNS messages just described, the interpretation of most of the fields varies. Update messages, denoted with Op Code = 5 in the DNS message header, are encoded as illustrated in Figure 3.7.

Contrast this format with that for non-Update DNS messages depicted in Figure 3.3. The message header is of the same structure as that of "normal" DNS messages, though the interpretation differs.

The **Zone** section identifies the DNS zone to be updated by this Update message.

The **Prerequisite** section enables the specification of conditions that must be satisfied in order to perform the update successfully. The condition and type of condition are determined by the value of each resource record-encoded parameter within the Prerequisite section. Table 3.4 defines how DNS Update prerequisites are interpreted based on the values of the Owner, Class, Type, and RData fields within the Prerequisites section.

The **Update** section contains the resource records to be added to or deleted from the zone using a similar encoding as used in the Prerequisite section as specified in Table 3.5.

The **Additional Data** section contains resource records related to this update; for example, TSIG signature records. Consider an example of an Update message received with the prerequisite and update fields encoded as specified in Table 3.6.

The Update section contents will only be considered if the prerequisite condition is met. In this case, the prerequisite condition is that the `host.ipamworldwide. com. IN DHCID H8349a+)3jELeA==ES1` record exists in the zone, that is, prerequisite type RRSet with matching owner, type, and RData (value dependent). If it does exist, then the `host.ipamworldwide.com. IN A 10.0.0.200` resource record from the Update section will be added to the zone. If not, the update will not be performed.

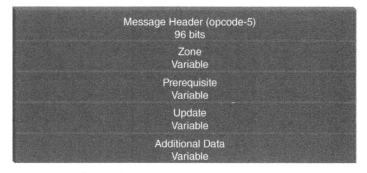

Figure 3.7. DNS Update Message Format

TABLE 3.4   DNS Prerequisite Interpretation

Owner	Class	Type	RData	Prerequisite Interpretation
Match	ANY (255)	ANY	Empty	The matching owner name is in use in this zone
Match	ANY (255)	Match	Empty	An RRSet with matching owner and type exists (value independent, i.e., any RData match)
Match	NONE (254)	ANY	Empty	The matching owner name is not in use in this zone
Match	NONE (254)	Match	Empty	An RRSet with matching owner and type does not exist in this zone
Match	Same as Zone Class	Match	Match	An RRSet with matching owner, type, and RData exists in this zone (value dependent, i.e., RData match)

This particular example illustrates how an entity can perform a dynamic update of DNS data upon assigning an IP address, in this case 10.0.0.200 to host.ipamworldwide.com. The DHCID record used in this example provides a hash of the host's hardware address receiving the IP address to uniquely identify the host. The prerequisite condition for updating the address record provides a means to assure that only the original holder of this A record can modify it, minimizing naming duplication and the probability of record hijacking.

DNS Extensions (EDNS0)   Thus far in our discussion of the DNS message header, one may observe that all code bits are assigned but one, and these days many hosts can process larger multipart UDP packets than the originally specified size limit

TABLE 3.5   DNS Update Interpretation

Owner	Class	Type	RData	Update Interpretation
Owner to add	Same as Zone Class	RR type	RR RData	Add this resource record(s) of the specified owner, type, and RData to the zone's RRSet
Owner to delete	ANY (255)	RR type	Empty	Delete the resource records of the specified owner and type
Owner to delete	ANY (255)	ANY	Empty	Delete all resource records of the specified owner name
Owner to delete	NONE (254)	RR type	RR RData	Delete the resource record(s) of the specified owner, type, and RData from the zone

TABLE 3.6   DNS Update Example

Field	Owner	Class	Type	RData
Prerequisite	host.ipamworldwide.com.	IN	DHCID	H8349a+)3jELeA==ES1
Update	host.ipamworldwide.com.	IN	A	10.0.0.200

of 512 bytes. As a result of these limitations, as well as the desire to add additional domain name label types, DNS extensions were defined in its current incarnation in RFC 6891 (20). EDNS0 essentially provides a signaling channel between a pair of DNS components (e.g., a recursive and an authoritative DNS server) in order to signal advanced feature support and compatibility.

RFC 6891 defines version 0 of extension mechanisms for DNS, hence the designation, EDNS0. The RFC addresses the above constraints by defining the following extensions:

- Extended label types enable the ability to define unique interpretations of labels in DNS messages. One such interpretation was to interpret label data as binary or bit-string labels. In practice, this turned out to be very difficult to implement across the Internet, so binary labels have since been reduced to experimental status. Nevertheless the capability exists with EDNS0 to extend domain label types, though their use is discouraged due to observed difficulties in their deployment on the Internet. As we discussed, the first two bits of the domain label uniquely identify the label as a length byte (first two bits = 00) or as a pointer (first two bits = 11). The extended label type has been assigned 01 as its first two bits.
- EDNS0 defines a pseudo-resource record, the OPT record (i.e., RRType = OPT). The OPT record is placed in the Additional section by the resolver or server to advertise its respective capabilities. The OPT resource record is used to advertise capabilities of the sender (client or server) to the recipient, and only one OPT record should be present.

The OPT pseudo-resource record is encoded as shown in Figure 3.8, enabling specification of the senders' UDP packet size and additional response code bits.

As a pseudo-resource record, the OPT record should never appear in a zone file. Thus, while the OPT record utilizes the same wire format as other resource records, the definition of standard fields has been modified to provide extension information. The NAME (aka Owner) field is zero for the OPT record. The TYPE is OPT, and the CLASS field indicates the maximum size of the sender's UDP payload. The 32-bit TTL field is divided into three fields:

- Extended Response Code – adds 8 bits to the 4-bit RCODE in the DNS message header to provide 12 bits total.

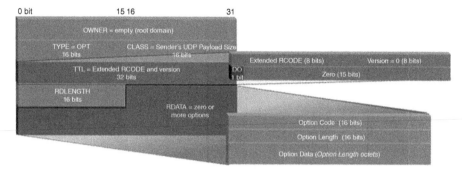

Figure 3.8. EDNS0 Format

- EDNS version number.
- Extended Header Flags – bit 0 is currrently defined as "DNSSEC OK" meaning that the querying server is capable of processing DNSSEC resource records. The remaining 15 bits of the extended header are currently reserved.

The RDLENGTH field indicates the length of the RDATA field, which consists of a set of zero or more options, each encoded with an option code, option length, and option value. Table 3.7 enumerates the options that have been allocated by IANA to date.

The Long-lived query* (LLQ; option code = 1) and Update Lease Life† (UL; option code = 2) are currently on-hold as RFCs and have not been officially published regarding these settings.

The name server identifier (NSID) option, defined with option code = 3, enables a resolver to request and a server to provide its identity as defined by the server administrator as its name, IP address, pseudorandom number, or other character string (configurable in BIND or PowerDNS using the `server-id` statement and in NSD and Knot using the `identity` parameter). This EDNS0 option is useful for debugging in environments where many servers share a common IP address, such as in deployments of anycast addressing or with load balancers.

Option codes 5–7 provide signaling of DNSSEC cryptographic algorithm support by a resolver. These options are only included in queries to inform the authoritative server of algorithm understanding, though they do not cause any change in server formulation of the response. Inclusion of the DNSSEC algorithm understood (DAU) option signals support of the specified digital signing algorithms. The DNSSEC hash understood (DHU) option indicates which Delegation Signer (DS) hash algorithms

---

* A long-lived query is a mechanism for a resolver to request receipt of notification of zone information changes; something like a DNSNOTIFY for clients.
† The Update Lease Life mechanism would enable a DHCP server to inform the DNS server within a DNS Update message of the corresponding client's lease length in seconds for new and renewed leases.

TABLE 3.7 Currently Assigned EDNS0 Options

Option Code					
Decimal	Hex	Name	Description	Status	Reference
0	0	Reserved	Reserved		RFC 6891
1	1	LLQ	Long-lived query	On-hold	http://files.dns-sd.org/draft-sekar-dns-llq.txt
2	2	UL	Update lease life	On-hold	http://files.dns-sd.org/draft-sekar-dns-llq.txt
3	3	NSID	Name server identifier	Standard	RFC 5001 (27)
4	4	Reserved	Reserved		http://www.iana.org/go/draft-cheshire-edns0-owner-option
5	5	DAU	DNSSEC algorithm understood	Standard	RFC 6975 (28)
6	6	DHU	DNSSEC hash understood	Standard	RFC 6975
7	7	N3U	NSEC3 hash understood	Standard	RFC 6975
8	8	edns-client-subnet	Stub resolver subnet address	Optional	RFC 7871 (29)
9	9	EDNS EXPIRE	SOA zone expiry	Optional	RFC 7314 (30)
10	A	COOKIE	DNS transaction security	Standard	RFC 7873
11	B	edns-tcp-keepalive	Idle timeout for TCP connections	Standard	RFC 7828 (31)
12	C	Padding	Artificial message size increase to hamper size-based correlation of encrypted DNS transactions	Standard	RFC 7830 (32)
13	D	CHAIN	Request DNSSEC validation chain	Standard	RFC 7901 (33)
14–65000	E–FDE8	Unassigned	Unassigned		
65001–65534	FDE9–FFFE	Reserved	Reserved for local or experimental use		RFC 6891
65535	FFFF	Reserved	Reserved		RFC 6891

are supported. And the NSEC3 hash understood (N3U) option signals support for NSEC3 hash algorithms.

Option 8, edns-client-subnet enables a stub resolver to indicate its IP subnet address. Some authoritative DNS servers supply or order resource records within a resulting resource record set based on the source IP address of the query in order to suggest a "closer" destination based on the IP address. However, with tiered resolvers where a stub resolver may query a recursive server which forward to a centralized caching server, the source IP address may be far removed from the stub resolver. This option enables the stub resolver to convey its IP subnet address to influence the response.

The EDNS EXPIRE option provides a means for a slave server querying another slave server to convey the correct zone expiry time. In some deployments, a set of slave servers may query a master server, while other slaves query one of these first-tier slaves. The expire time field of the SOA record is a relative time value and it can be erroneously interpreted as extended by each slave query tier.

The COOKIE option enables a lightweight transaction security mechanism which we will discuss in Chapter 7. The EDNS keepalive option signals willing-ness to keep an idle TCP connection open to reduce the overhead of connection reestablishment for subsequent transactions. Option 13, CHAIN, requests the com-plete DNSSEC validation path to a specified point in the domain tree to eliminate the resolver from having to subsequently issue queries for this information. We'll discuss the DNSSEC validation process and how this option can improve validation efficiency in Chapter 8.

THE DNS RESOLUTION PROCESS REVISITED

Now that we've explored the details of the DNS protocol, let's take a deeper dive into the DNS resolution process to illustrate the query processing with respect to particular DNS protocol parameters. Consider Figure 3.9, which is identical to Figure 2.2, repro-duced here for convenience. The recursive server accepts recursive queries directly from client (stub) resolvers and performs the following steps to obtain the answer to the query on behalf of the resolver.

1. The stub resolver initiates a query to the recursive DNS server. The resolver knows which DNS server to query based on configuration via manual entry or via DHCP. Figure 3.10 illustrates an example DNS packet issued by the resolver client. The client's IP address is 10.10.0.23 while the internal IP address of the recursive server is 10.20.5.100 as configured in the resolver, for example, via DHCP. The resolver uses a random source UDP port, 12510 in this case, and the standard DNS destination port, 53. The resolver sets the transaction ID field to a random value as well, 37321 here. Flag set-tings indicate this as a standard query where recursion services are desired,

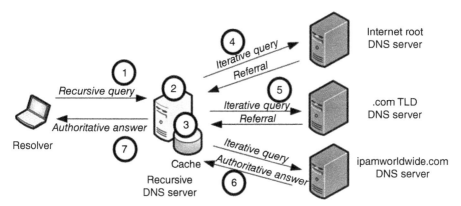

Figure 3.9. Recursive and Iterative Queries in Name Resolution

requesting the recursive server to perform all queries necessary to resolve the query. Finally, the single question, here worded as a question, though in an actual packet would contain just the Qname (www.example.com), Qclass (IN), and Qtype (A).

2. The queried recursive DNS server will first search its configured data files. That is, the DNS server is typically configured with configuration and resource record information for which it is authoritative. If the answer is found, it is returned to the resolver and the process stops. If not and the server is configured to forward queries for this zone or for all queries, it does so and awaits a response.

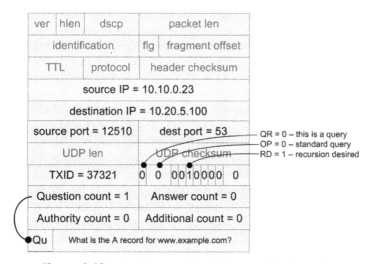

Figure 3.10. Resolver-Issued DNS Query Packet Example

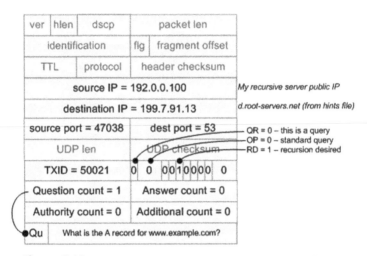

Figure 3.11.   Recursive Server Query to an Internet Root Server

3. If the queried recursive DNS server is not authoritative for the queried domain nor configured to forward this query, it will access its cache to seek the answer to the query.

4. If the queried recursive DNS server cannot locate the full query name information in cache, it will then attempt to locate the information via another DNS server using any cached partial domain information or by accessing its configured root hints file to begin traversing down the domain hierarchy to a DNS server that can provide an answer to the query.

   In our example depicted in Figure 3.11, a root name server is selected from the root hints file, d.root-servers.net with IP address 199.7.91.13. The query is issued using the DNS server's public IP address, 192.0.0.100, as its source address for proper Internet routing on well-known DNS destination UDP port 53. A random source UDP port and transaction ID should be populated along with header bits indicating this is a query of the standard variety. Some recursive servers will set the recursion desired bit in the header, but this will be ignored by the root and TLD servers, and should be ignored by all external name servers including yours (by disabling recursion).

   The root name server response refers the querying recursive DNS server to the name servers that are authoritative for the com domain as illustrated in Figure 3.12. The referral response is directed to the recursive server IP address and the UDP port it used in its query. The transaction ID is copied as well. The DNS header indicates no answer is provided to the query, but in this example, two records each are included in the Authority and Additional sections. The authority records provide the name server (NS) records of the servers authoritative for the .com domain. The Additional section contains

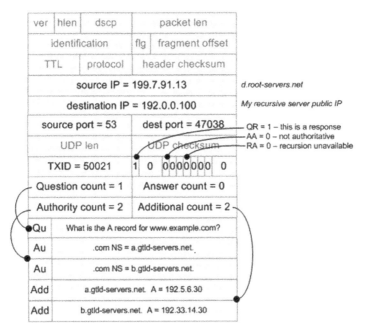

Figure 3.12. Example Root Server Referral Response

corresponding A (and/or AAAA) records that map the name server names to IP addresses. These additional records are the "glue" records to map the child domain's name server domain names to IP addresses to which further queries can be directed.

5. Having received this referral response, the recursive server selects one of the .com name server IP addresses and issues a query such as that shown in Figure 3.13. The answer received will also be a referral to the name server that is authoritative for example.com. per Figure 3.14, and so on down the tree. Note that each query should use random TXID values and monotonically increasing values should NOT be used for each successive transaction. This just makes it that much easier for attackers to spoof responses.

   This referral response is very similar to that received from the root servers though for one layer down in the domain tree.

6. Upon receipt of the referral from the TLD server indicating two name servers and corresponding IP addresses that are delegated to example.com, the recursive server prepares its next iterative query, which looks very similar to those sent to the root and TLD servers as shown in Figure 3.15.

   The response, illustrated in Figure 3.16, includes the answer with one answer record, and an authoritative answer at that, with the AA flag set in the DNS header. The Authority and Additional sections also contain the name

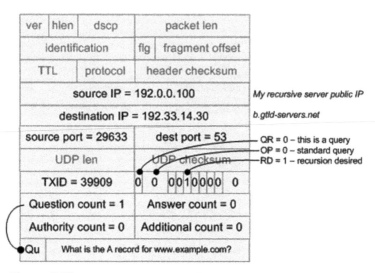

Figure 3.13. Recursive Server Iterative Query to a .com Name Server

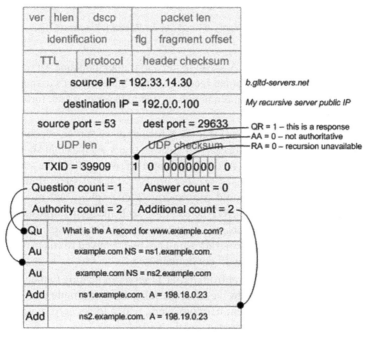

Figure 3.14. TLD Server Referral

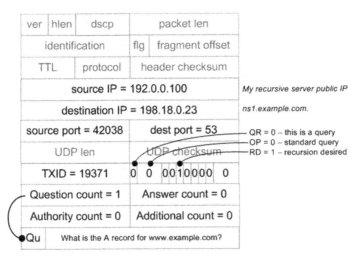

ver	hlen	dscp	packet len	
identification		flg	fragment offset	
TTL	protocol	header checksum		
source IP = 192.0.0.100				*My recursive server public IP*
destination IP = 198.18.0.23				*ns1.example.com.*
source port = 42038		dest port = 53		QR = 0 – this is a query
UDP len		UDP checksum		OP = 0 – standard query RD = 1 – recursion desired
TXID = 19371		0 0 0010000 0		
Question count = 1		Answer count = 0		
Authority count = 0		Additional count = 0		
Qu	What is the A record for www.example.com?			

Figure 3.15. Recursive Query to an example.com Name Server

ver	hlen	dscp	packet len
identification		flg	fragment offset
TTL	protocol	header checksum	
source IP = 198.18.0.23			
destination IP = 192.0.0.100			
source port = 53		dest port = 42038	
UDP len		UDP checksum	
TXID = 19371		1 0 1000000 0	
Question count = 1		Answer count = 1	
Authority count = 2		Additional count = 2	
Qu	What is the A record for www.example.com?		
Ans	www.example.com. A = 192.0.2.54		
Au	example.com NS = ns1.example.com.		
Au	example.com NS = ns2.example.com		
Add	ns1.example.com. A = 198.18.0.23		
Add	ns2.example.com. A = 198.19.0.23		

ns1.example.com

Recursive server IP

Figure 3.16. Authoritative Answer

server and address records for those servers authoritative for the example.com zone. The recursive server updates its cache generally not only with the ultimate answer for the specific query, but also for the contents of the Authority and Additional sections for the answer and referral messages received in the process. When the same or another stub resolver queries for another domain with the .com domain subtree, the recursive server can utilize its cache to query one of the .com name servers directly without needing to query the root server. If an answer cannot be found, the recursive server will also cache this "negative" information as well for use in responding to similar queries. Such negative cache entries expire from cache based on the zone's negative cache TTL parameter within its SOA record.

7. When the answer is received, the recursive DNS server will provide the answer to the stub resolver per Figure 3.17 and also update its cache and the process ends. Notice that the query answer maps the UDP ports and DNS transaction ID initially provided by the stub resolver upon its initial query.

The stub resolver may be configured to cache this information as well. In this case, only the answer is cached, not other domain tree nodes, because the stub resolver always queries its configured name servers. But for users

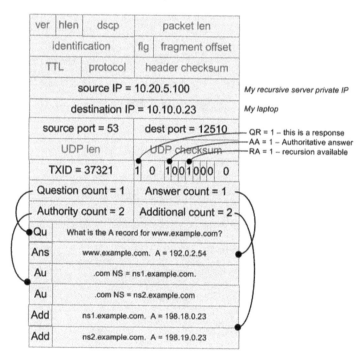

Figure 3.17. Query Answer Example

frequently visiting common sites, the stub resolver cache can greatly improve application response time.

The key parameters used by the recursive server to match responses with outstanding queries consist of the following:

o Source IP address matches the IP address to which the query was sent and the destination IP address matches the address that was used as the outbound source address.

o Destination UDP (or TCP) port matches the source port number used on the outbound query. The source port should also match the outbound destination port, though this is almost always port 53, the well-known DNS port.

o The DNS header transaction identifier matches that used in the outbound query.

o The question (name, class, and type) match that posed in the outbound query and the answer name matches. If character case matching is supported, the case of the Qname must match the Answer in the response.

o The Authority (name server) name falls within a common domain branch (e.g., example.com) as the Answer section name.

Should all of these parameters match, the recursive server will consider the answer valid and will continue processing by adding the record to cache and providing the answer to the resolver and future resolver queries for the duration of the answer record's TTL. In Chapter 8, we'll analyze attacks on recursive servers, cache poisoning attacks in particular, and how such attacks can infiltrate these servers.

DNS Resolution Privacy Extension

In the prior resolution example, note how the Question section always contained the full FQDN of the query name for which an answer was sought. In our example, three separate packets traversed the Internet with this query FQDN from the recursive server IP address. Someone snooping query traffic or server cache may be able to infer information about clients based on what they are querying. The root and TLD name servers will never provide a complete answer to a typical query since they are delegation servers by design. Thus, querying for the FQDN provides the root and TLD servers (and any other ancestor domains in a long-labeled domain name) with this query information, which in reality is "too much information."

To address this potential privacy concern, RFC 7816 (34) was published to specify query name minimization. Though an experimental RFC, it stipulates that the query name specified in queries to root name servers should only include the TLD for which a referral is sought. The root servers will never answer with any further detail than that in any case. Likewise, when querying the TLD servers, only the next layer down should be included as the query name. Hence for an end user query for pc52.dev.ipamworldwide.com would result in a query to the root server for "com." and to the .com server for "ipamworldwide.com." and so on down the domain tree.

The intended effect is to minimize provision of "TMI" for Internet name servers and possible eavesdroppers seeking query information for general use or to classify user browsing or behavior.

SUMMARY

This chapter has discussed details of the DNS protocol, including message types and formats. Equipped with this foundation, we will now move on to introduce vulnerabilities and risks of the DNS protocol and your overall DNS infrastructure.

4

DNS VULNERABILITIES

INTRODUCTION

This chapter serves as an introduction to the threats and vulnerabilities for your DNS infrastructure. These threats span those to which all network components are vulnerable, to those specific to the DNS protocol and server implementations, to broader network vulnerabilities thanks to the ubiquitous availability of DNS services. We'll first build on our foundation of DNS architecture to define a trust model which is helpful for scoping the potential sources of vulnerabilities. We'll then briefly introduce each vulnerability with an example attack vector in this chapter. We'll delve more deeply into each vulnerability and associated attack methods and mitigation strategies in subsequent chapters.

DNS DATA SECURITY

As a network service, DNS services run on a physical or virtual DNS server. Hence, securing physical server access, the operating system, and the DNS server implementation are critical considerations. For example, specific DNS server implementations may contain certain vulnerabilities. As with your other network services, securing access and monitoring of operating system and associated software vulnerabilities is a

DNS Security Management, First Edition. Michael Dooley and Timothy Rooney.
© 2017 by The Institute of Electrical and Electronic Engineers, Inc. Published 2017 by John Wiley & Sons, Inc.

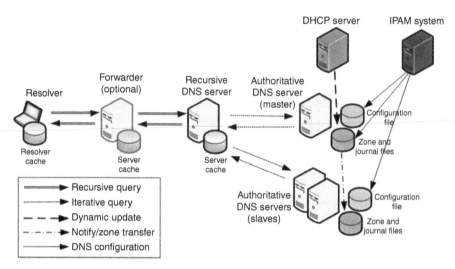

Figure 4.1. Basic DNS Data Stores and Update Sources

fundamental operational process. Beyond server security considerations, it is instructive in discussing DNS security vulnerabilities to consider the DNS data sources and data flows, depicted in Figure 1.2, repeated here for convenience in Figure 4.1.

Starting in the upper right-hand corner of the figure, DNS servers are initially configured with configuration and zone information. This configuration step may be performed using a text editor, an IPAM (IP address management) system, or other vendor-provided DNS management system. Configuration of server parameters and associated zones is required to initialize the server for operation within its respective role and with your namespace's resolution data.

This configuration is typically defined within a configuration text file, for example, named.conf, nsd.conf, knot.conf, or pdns.conf, though the latter pair support scripting and/or database backends. Associated zone repositories define the resolution data for each authoritative zone on the master server. While a server may be master for some zones and slave for others, it's generally simpler to designate one server among a set be the master for a set of zones while deploying multiple slave servers for these zones. We'll use the master server terminology in assessing the vulnerability of a particular zone's information.

Configuration of slave servers requires creation of the configuration file only, which defines the server's configuration parameters and its authority for particular zones. Slave servers transfer zone information in the form of resource records from corresponding master servers. Deleting zones or adding new ones typically requires updating of the configuration file deployed on each slave DNS server to define its universe of authoritative zones. However, PowerDNS and ISC BIND have implemented super masters and catalog zones, respectively, to enable zone additions and deletions to require only updating of the master. The master will replicate zone adds and deletes in addition to resource record changes within the standard zone transfer process.

Zone information may be updated by external sources as well, particularly DHCP servers. Dynamic updates can be accepted for clients obtaining dynamic IP addresses requiring DNS updating of address-to-name mappings. These updates will typically originate from the DHCP server assigning the address and will be directed to the server acting as master for the given zone. The master will add the update to its zone repository and may then notify its slaves of the update, who may request an incremental zone transfer to capture the updated zone information.

Beyond the configuration information and zone files, the third information repository within a DNS server is its cache. Cache information is accumulated through the query resolution process. As query answers are sought and received, corresponding answers are cached by the server. The cached information may be obtained not only from the Answer section of the DNS protocol message, but also from the Authority and Additional sections. These sections supply authoritative server information and information purportedly supplemental to the Answer. This information may include the authoritative servers for the relevant zone and other information related to the query (e.g., the A/AAAA "glue" record for an NS query). Please refer to Chapter 2 if you need to review the query resolution process.

DNS INFORMATION TRUST MODEL

The *DNS information trust model* refers here to how DNS information flows among these components of the DNS system, is accepted as from trusted sources, and is stored on a given component. Each component's role defines what information and from what sources it requires to perform its function. Leveraging this fundamental operation, attackers may attempt to manipulate this data at rest or in motion for nefarious purposes as we'll discuss in the remainder of this book. Of course, the DNS security administrator must apply controls to delineate what information received by certain other components can be trusted for use and storage as well as any validation and authentication mechanisms to improve trustworthiness.

From the client stub resolver perspective, the resolver trusts its configuration of which DNS server(s) to query, its resolver cache, and the availability and integrity of queried recursive servers (and intervening network) to provide answers to DNS queries. Should any trusted data source be corrupted, the resolver could inadvertently redirect the user application to an inappropriate destination. If queried servers fail to answer queries, the DNS service will be rendered effectively disabled, leaving users unable to navigate the web.

A forwarding DNS server, if used, relies on its cache, its forwarder configuration, and on the availability and integrity of the recursive DNS servers to which it forwards queries. If authoritative for certain zones, the server relies on the accurate configuration of respective zone data, which implies accurate, secure storage and data encryption and verification if configured remotely.

Our next component, continuing from left to right, the recursive server, trusts its configuration (e.g., root hints file contents), its cache and the various DNS servers it queries internally on the enterprise network and/or externally on the Internet. It relies not only on accurate responses from authoritative DNS servers but on other DNS servers which provide referrals to locate DNS servers authoritative for the domain in question. Referral answers are those provided by the Internet root servers as well as top level domain (TLD) servers, but referrals may also be provided by other DNS servers operated by other organizations or by DNS hosting providers.

Authoritative DNS servers are so called given that they are purportedly configured and operated by or on behalf of the administrator of a given DNS domain who is responsible for the information published on these servers. Resolvers querying hostnames within the domain of the authoritative server trust the referral servers to direct them to the proper set of authoritative servers, and resolvers trust the queried authoritative server to respond with accurate information, where *accurate* means *as published by the domain administrator*.

As a domain administrator, you'll publish DNS resolution data within your zones to enable your users to resolve domain names for resources available from within your organization. We'll define resolution data for internal users to access internal resources ingeniously enough, as your *internal namespace*. You'll also need to publish resolution data for servers you desire to make available via the Internet. This set of servers or resources in general is typically an independent set of servers with corresponding IP addressing. As such, we'll refer to this resolution data as defining your *external namespace*.

You can publish resolution information on your authoritative DNS servers via a variety of sources as we've seen, including manually edited text files, inter-server transfers and updates, and/or using a DNS or IPAM system. Inter-server transfers refer to master–slave replication, while updates may originate from other DNS servers, DHCP servers, other systems, or even end user devices if permitted by administrators.

DNS Information Sources

The DNS trust model provides us the per component perspective on information flow among components and inherent trust relationships within the resolution process. Examining the sources of each component's information can help us understand how each is potentially configured both in terms of base configuration, such as for what domains a particular server is authoritative, whether it should answer queries from given addresses, and actual DNS resolution data communicated via the DNS protocol.

Table 4.1 summarizes potential sources of this information, respectively. Referral DNS servers include root servers, TLD servers, and other *ancestor* (e.g., parent, grandparent) DNS servers within the domain tree that refer recursive DNS servers

TABLE 4.1   DNS Information Sources

Component	Configuration Data Sources	DNS Data Sources
Client resolver	• Configuration or properties file • DHCP/PPP server-provided parameters	• Resolver cache (if equipped) • Recursive DNS servers
Forwarding DNS server	• Configuration or properties file • DNS control channel	• Server cache • Recursive DNS servers
Recursive DNS server	• Configuration or properties file • DNS control channel • Hints file of root DNS servers	• Server cache • Referral DNS servers • Authoritative DNS servers
Referral DNS servers	• Domain level DNS administrators and tools	• Child domain administrators provision of their DNS servers' names/IP addresses (NS/glue records)
Authoritative DNS server for internal or external namespace	• Configuration file or properties • Zone file(s) • DNS control channel	• DNS administrator via text editor, DNS GUI or IPAM system • DDNS updates to zone files • Zone transfers

down the tree ultimately to the DNS servers authoritative for the FQDN queried. These servers are administered by external organizations with IANA managing the Internet root zone, and TLD administrators managing corresponding DNS servers to publish proper mapping of child domains to corresponding authoritative or referral DNS servers.

DNS Risks

The viability of your DNS services to your constituents requires your diligent management of the risks or vulnerabilities to the integrity and availability of these DNS services. Risks include not only security risks, for example, malicious attacks, but also deployment design and management procedures. Inadequate redundancy design could render DNS services unavailable to your users with the outage of a network segment or one or more DNS servers. Natural disasters or human error can likewise render DNS offline.

Attackers may target DNS services in and of themselves in order to stifle communications or to steer unwitting end users to imposter web servers or other destinations. Alternatively, DNS may serve as a facilitator for use with the scope of a broader network attack. Just as DNS enables users to connect to websites by resolving text-based destinations to IP addresses, it enables attacker malware to locate command and control centers or to tunnel information through firewalls. DNS by its nature also openly publishes potentially useful information about networks, host names, and IP addresses for would-be attackers.

We'll first examine design risks, then malicious attacks on DNS infrastructure consisting of DNS servers within your organization's control and those on the Internet used within the process of name resolution. Then we will discuss more broadly targeted network attacks that leverage the DNS. As we shall see, given this broad and diverse set of attack vectors, no single mitigation technology can effectively combat them all; a comprehensive defense in depth DNS security strategy is necessary to defend against them collectively.

DNS INFRASTRUCTURE RISKS AND ATTACKS

By DNS infrastructure, we refer to the DNS components we've discussed: the servers themselves, the DNS resolvers which query for address lookups on behalf of user applications on end user devices, the integrity of DNS information, and the collective DNS service of resolving hostnames on behalf of resolvers. We'll introduce the key risks and vulnerabilities in the remainder of this chapter, and follow with respective mitigation strategies in ensuing chapters.

DNS Service Availability

Your users rely on DNS for the basic network functions. While not within the realm of attacks per se, proper deployment and management of DNS infrastructure is critical to providing always available DNS with adequate resolution performance for your constituents. Even the most secure networks can be ineffective if DNS services are unavailable. Risks to availability of DNS services include

- Inadequate DNS capacity due to too few servers deployed and/or servers deployed with inadequate processing or memory.
- Unavailability of DNS services due to network unreachability, that is, poor network placement of DNS servers.
- The failure of a DNS server due to hardware failure, power failure, natural disaster, or human error can cause unreachability due to a server or subnetwork failure and can increase the load on other DNS servers authoritative for the same zones.
- Failure to segment servers by "role" (i.e., authoritative vs. recursive) can overload servers and expose them to multiple attack vectors.

Hardware/OS Attacks

As with all network servers, vulnerabilities within the server operating system and applications may be exploited by attackers in order to severely hamper or crash the server. These attacks can be of the following forms:

- Hardware – Physical access to DNS servers enables the attacker to unplug, disconnect, or physically remove the server, literally removing the server from service, thereby reducing the availability of the DNS service and possible capture of configuration information. Physical removal of a server affords the attacker an opportunity to hack the server for zone information, DNS infrastructure information and possibly private keys used for DNSSEC.
- Operating system attacks – An attacker may attempt to gain local or remote console access to the server by hacking passwords or overflowing the code execution stack or buffer. In general, an attack may exploit a known vulnerability of the operating system or kernel software running on the server.
- DNS service attacks – An attacker may attempt to exploit a known vulnerability for a given vendor and version of DNS server software running on the victim server to shut it down or otherwise corrupt and/or disrupt service.
- Control channel attack –The DNS server control channel provided in most implementations provides a convenient mechanism to remotely control the DNS server, such as stopping/halting the server's DNS software (e.g., "named," "nsd"), reloading a zone, and more. Such power may entice an attacker to attempt to access the control channel to perform nefarious functions such as stopping the DNS service thereby denying DNS service to querying servers and resolvers.

These types of attacks potentially affect all DNS components as illustrated in Table 4.2.

DNS Service Denial

The all too familiar denial of service (DoS) or distributed DoS (DDoS) attack is invoked by an attacker with the intent to flood the DNS server with bogus DNS requests or other irrelevant packets, overwhelming its ability to process legitimate DNS queries. From the DNS server's perspective, it merely attempts to process each packet as it is received. As the volume of bogus packets is intensified beyond the query response rate supported by the server, the proportion of legitimate queries processed lessens and DNS resolution services capacity drops precipitously to only that small proportion that is processed. Ultimately such an attack may crash the server altogether.

TABLE 4.2   Server and OS Attacks

Attack Type	Target Component	Potential Impact
Server and operating system attacks	Stub resolver	Infiltration of the stub resolver or system could disable resolution or redirect resolutions to an attacker recursive DNS server, directing client applications to attacker-defined destinations potentially compromising sensitive information
	Forwarding DNS server	An attacker could shutdown resolution or query forwarding or reconfigure forwarders to point to attacker DNS servers
	Recursive DNS server	An attacker can shutdown recursive services or possibly replace the hints file to direct noncached queries to the attacker's "root" servers for full resolution control
	Referral DNS server	Could impact availability of the DNS service or resolution of child domain subtrees if the attacker modifies the referral zone file; the higher in the domain tree the broader this impact of manipulating the zone file contents
	Authoritative DNS server for internal namespace	Could impact availability of DNS service or resolution data integrity if attacker manipulates zone files for the domain and subdomains
	Authoritative DNS server for external namespace	Could impact availability of DNS service or resolution data integrity if attacker manipulates zone files for the domain and subdomains

As illustrated in the crude query processing chart on the right side of Figure 4.2, prior to the attack, the server is receiving queries well within its capacity to respond. The attacker substantially increases the volume of queries in an effort to inundate the server and reduce or eliminate DNS resolution.

The types of DoS/DDoS attacks may be in the form of the following:

- DNS query flood – the attacker issues a large number of DNS queries beyond which it has capacity to resolve.

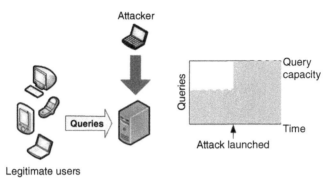

Figure 4.2. Denial of Service Attack

- UDP packet flood – an attacker may issue large numbers of UDP packets using random UDP destination port numbers, forcing the server to respond with an ICMP Destination Unreachable message for each.
- TCP SYN attack – while DNS typically utilizes UDP, TCP is permitted and the SYN attack involves the attacker opening a TCP connection by sending the TCP SYN message from varying source IP addresses and/or ports, then ignoring the SYN-ACK thereby not completing connection establishment with this third message of the three-way handshake. While awaiting each ACK, the server keeps the half-open TCP connection pending, and ultimately depletes its capacity for TCP connections.
- ICMP flood – the attacker issues a constant stream of ICMP packets to the server which uselessly occupies its processing capabilities.

Distributed Denial of Service A variant of this type of attack is the use of multiple distributed attack points and is referred to as a distributed denial of service (DDoS) attack. The intent is the same, though the scale is much larger, with multiple attack origination points as illustrated in Figure 4.3.

Attackers can enlist others to manually conduct an attack on a target simultaneously. However, in many cases, the use of bots installed on other computers within the enterprise or on the Internet can be enlisted to join in the attack.

This was basically what occurred on October 21, 2016 during the DDoS attack on DynDNS. Based on a statement about the attack from Dyn (35), DNS packets from tens of millions of IP addresses associated with the Mirai botnet barraged Dyn's DNS infrastructure. The Mirai malware had infected over 100,000 devices, predominantly non-person entities (NPEs) otherwise known as Internet of Things (IoT) devices, and enlisted these devices in the attacks.

Bogus Domain Queries This attack attempts service denial through the flooding of a recursive server with queries for bogus domain names. This causes

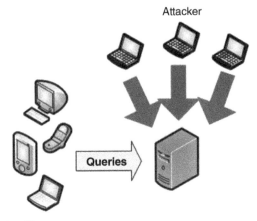

Figure 4.3.   Distributed Denial of Service Attack

the server to initiate a "wild goose chase" and utilize resources to futilely locate the authoritative server within the domain tree as shown in Figure 4.4.

In addition to processing a high volume of such queries as in a typical DoS attack, the recursive server expends resources iterating queries to name servers within the domain tree in an attempt to identify the authoritative servers for each bogus domain. Ultimately, query errors will be returned for lame delegations or NXDOMAIN responses but the sheer volume of such pending queries can inhibit its processing of legitimate queries.

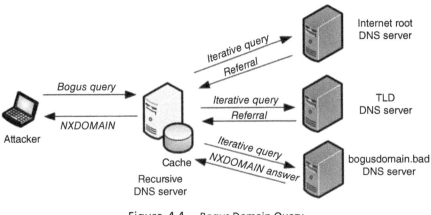

Figure 4.4.   Bogus Domain Query

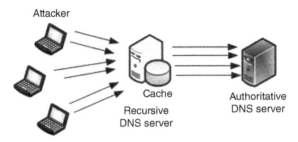

Figure 4.5. PRSD Attack

Pseudorandom Subdomain Attacks

A variant of the generic bogus domain query attack focuses queries on a given domain served by a set of authoritative servers. This attack vector has been shown to impact not only the authoritative servers but recursive servers awaiting responses from these authoritative servers. This attack, called a pseudorandom subdomain (PRSD) attack features an attacker launching a large number of queries containing pseudorandom subdomains of a target domain, let's say example.com. Thus, an attacker queries for names like iopqewf.example.com, a84fj.example.com in large volumes. The large volume of queries can inundate the DNS servers authoritative for the example.com domain, thereby denying service.

Worse still, the ripple effect on the recursive server(s) to which the queries have been launched, for example, the attacker's ISP's DNS servers, can be debilitating. Once the authoritative servers have essentially crashed, the recursive server continues processing queries. As the number of outstanding unanswered queries grows, the ability of the recursive servers to handle new legitimate queries diminishes, thereby reducing or even denying recursive DNS services for the ISP's customers.

Figure 4.5 shows the basic attack flow. An attacker may enlist a botnet formed from a collection of malware-infected residential devices of a given ISP. When the attack ensues, attacker resolvers flood the ISP's recursive server with queries requesting resolution for PRSD names beneath the target domain as discussed above. As the queries mount against the authoritative DNS servers, legitimate queries are drowned out and ultimately the authoritative servers may crash. Meanwhile, as the recursive DNS servers continue to launch queries, they may exhaust their resources awaiting responses on unanswered queries to the target authoritative DNS servers. This service denial not only impacts the target domain's DNS servers but the ISP's recursive servers. Table 4.3 summarizes denial of service attack types and potential impacts.

Cache Poisoning Style Attacks

DNS resolvers and recursive caching servers maintain a cache of resolved resource records to improve resolution performance as described earlier. If an attacker

TABLE 4.3   Denial of Service Attack Summary

Attack Type	Target Component	Potential Impact
DoS and DDoS	Stub resolver	Slow down or inability of the host on which the resolver is running to support full networking capabilities
	Forwarding DNS server	Slow down or inability of all internal users or customers configured to use this forwarding DNS server for resolving domain names
	Recursive DNS server	Slow down or inability of all internal users or customers configured to use this recursive DNS server for resolving domain names
	Referral DNS server	Slow down or inability of all Internet users to resolve child domains of the attacked referral DNS server, e.g., attacking a TLD set of name servers to deny resolution of the TLD's domain branch (all descendant domains)
	Authoritative DNS server for internal namespace	Slow down or inability of internal users to resolve internal domain names, e.g., intranet
	Authoritative DNS server for external namespace	Slow down or inability of all Internet users to resolve your external (public) namespace
Bogus domain queries	Stub resolver	Should have no impact if the stub resolver is properly configured to ignore incoming DNS queries
	Forwarding DNS server	Slow down or inability of all internal users or customers configured to use this forwarding DNS server for resolving domain names
	Recursive DNS server	Slow down or inability of all internal users or customers configured to use this recursive DNS server for resolving domain names
	Referral DNS server	Should have no impact if the server is properly configured to disallow recursion
	Authoritative DNS server for internal namespace	Should have no impact if the server is properly configured to disallow recursion

TABLE 4.3 *(Continued)*

Attack Type	Target Component	Potential Impact
	Authoritative DNS server for external namespace	Should have no impact if the server is properly configured to disallow recursion
Pseudorandom subdomain (PRSD) attack	Stub resolver	Should have no impact if the stub resolver is properly configured to ignore incoming DNS queries
	Forwarding DNS server	Slow down or inability of all internal users or customers configured to use this forwarding DNS server for resolving domain names
	Recursive DNS server	Slow down or inability of all internal users or customers configured to use this recursive DNS server for resolving domain names
	Referral DNS server	Should have no impact if the server is properly configured to disallow recursion
	Authoritative DNS server for internal namespace	If your domain is targeted, users will experience a slow down or inability to resolve domain names from your DNS servers; otherwise, it should have no impact
	Authoritative DNS server for external namespace	If your domain is targeted, users will experience a slow down or inability to resolve domain names from your DNS servers; otherwise, it should have no impact

succeeds in corrupting a recursive server's cache, the corrupted information may be provided to several users requesting the same or similar domain name information. Corrupting the cache requires an attacker to provide a seemingly legitimate query answer albeit with falsified resolution information in part or in total.

These types of attacks are generally conducted as shown in Figure 4.6 where an attacker appears to the recursive server as the legitimate authoritative server to which it issued the query. In the various forms of this attack, ultimately the attacker attempts to corrupt the cache of the recursive server, for example, by pointing the resolution of a legitimate and even popular web or server address to a server operated by the attacker. The falsified resolution data is returned to the originator of the query and is also returned to other resolvers querying for this information while the corrupted

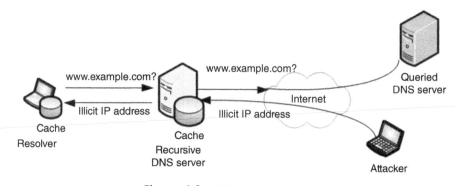

Figure 4.6.   DNS Cache Poisoning

information resides in cache, that is, for the duration of the TTL. This has the effect of hijacking potentially several resolvers and hence applications to incorrect destinations, for example, websites.

To corrupt the cache, the DNS query response from the attacker must reach the server before the legitimate response and map to an outstanding query for which the recursive server is awaiting a response. The server will map a received answer to a previously issued query by matching the following fields in the response:

- The source IP address of the response maps to the destination IP address of the query and the destination IP address matches the address of this server.
- The destination port of the response with the source port of the query and the answer's source port is 53.
- The DNS transaction ID within the DNS header matches on both the query and the response.
- The DNS Qname, Qclass, and Qtype in the question section matches on both the query and the response.
- The domain names in the Authority and Additional sections of the response must fall within the same domain branch as the Qname. This is known as the bailiwick check.

Consider a recursive DNS server that receives two matching responses as Figure 4.7. Which is authentic? The key parameters that the recursive server is seeking match on each answer to the outstanding query. The DNS server will accept the first matching answer it receives, cache it, and respond to the stub resolver. If the attacker can match the parameters with an answer that arrives before the legitimate answer from the authoritative name server, he or she will have succeeded in poisoning the cache with an answer that will be provided to other clients querying similar information.

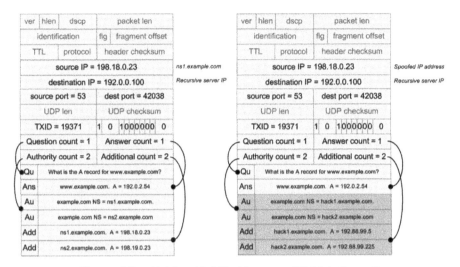

Figure 4.7. Valid and Falsified Response; Which is Correct?

Consider a recursive name server receiving such an answer message that matched on the IP addresses, UDP port numbers, transaction ID, query parameters, and bailiwick validation, but with falsified resource record information in one or more record sections. Most cache poisoning attacks modify the answer itself, pointing www.example.com to an IPv4 address operated by the attacker. The Kaminsky attack actually manipulates the names or addresses with records of the Authority or Additional sections, and may even provide an accurate answer in the Answer section. We'll discuss this and the other common forms of cache poisoning attacks in more detail in Chapter 8. Meanwhile, Table 4.4 summarizes the impacts of these cache poisoning attacks.

Authoritative Poisoning

Cache poisoning attacks attempt to corrupt DNS information cached within resolvers, forwarding servers, and recursive servers. Other forms of DNS information attacks attempt to corrupt that DNS information published within authoritative DNS servers. Unlike cached information which eventually times out, corruption of authoritative information could persist for a lengthy time period until detected and corrected.

- Dynamic updates – An attacker may attempt to inject or modify data in a DNS zone by attempting to issue a DNS Update message to the DNS server. This type of attack could manipulate resolution data, redirecting resolutions from clients for the intended destination to an attacker-specified destination.

TABLE 4.4   Cache Poisoning Attack Summary

Attack Type	Target Component	Potential Impact
Cache poisoning	Stub resolver	Infiltration of the stub resolver cache would enable redirection of client applications to attacker-defined destinations potentially compromising sensitive information for this client
	Forwarding DNS server	A potential target for cache poisoning as cache manipulation can affect multiple users, those who query the affected forwarding DNS server for domain information that has been manipulated
	Recursive DNS server	A prime target for cache poisoning as cache manipulation can affect multiple users, those who query the affected recursive server for domain information that has been manipulated
	Referral DNS server	Should have no impact if the server is properly configured to disallow recursion (and therefore caching)
	Authoritative DNS server for internal namespace	Should have no impact if the server is properly configured to disallow recursion (and therefore caching)
	Authoritative DNS server for external namespace	Should have no impact if the server is properly configured to disallow recursion (and therefore caching)

- Server configuration – An attacker may attempt to gain access to the physical server running the DNS service. One of the many actions an attacker can take upon gaining access is to edit DNS files residing on the system to manipulate resolution data. Assuming the infiltrated server is a DNS master for its configured zones, modified DNS zone data will be automatically conveyed to zone slaves to appear fully authoritative.
- Among other attacker steps beyond being able to manipulate configuration and zone information, an attack of this type could enable the use of the server as a stepping stone to other targets, especially if this server is trusted internally.
- Configuration errors – While typically not malicious (though most attacks are initiated from internal sources), misconfiguring the DNS service and/or zone information may lead to improper resolution or server behavior.

Table 4.5 summarizes the potential impacts of authoritative poisoning.

TABLE 4.5   Authoritative Poisoning Attacks

Attack Type	Target Component	Potential Impact
Authoritative poisoning	Stub resolver	Should have no impact as the resolver is not authoritative for any zone information
	Forwarding DNS server	To the extent that the forwarding server is also authoritative for one or more zones, an attack on a forwarding server could inhibit or redirect certain or all query resolutions
	Recursive DNS server	Should have no impact if the server is properly configured as a recursive-only server without being authoritative for any zone data
	Referral DNS server	An attacker infiltrating a DNS server could redirect child domain delegations to attacker name servers; the higher in the domain tree, the larger the affected Internet namespace
	Authoritative DNS server for internal namespace	An attack could inhibit resolution for certain or all records, subdomain delegations, or redirect internal zones and namespace erroneously or to an attacker destination
	Authoritative DNS server for external namespace	An attack could rename or delete key external domain names (e.g., www) and subdomains

Resolver Redirection Attacks

The stub resolver on a client device must be initialized with at least one DNS server IP address to which DNS queries can be issued. This IP address is the destination address for all DNS queries originating from the client. Other resolver configuration information such as domains suffixes may also be defined as we discussed in Chapter 2. The resolver configuration may be performed manually by hard coding the DNS server IP address in the TCP/IP stack, or automatically via DHCP or PPP. If an attacker can redirect recursive queries to a server under their control, any and all resolution information can be corrupted at will.

- Corruption through DHCP/PPP – This type of attack seeks to redirect the resolver from the legitimate recursive DNS server to an attacker's DNS server to poison the resolver with malicious DNS query answers. Manipulation of client configuration obtained through DHCP or PPP would generally

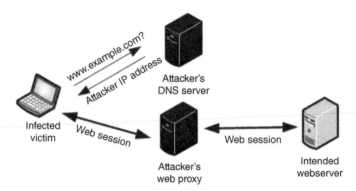

Figure 4.8.   Resolver Infiltration for Man in the Middle Attack

require the provision of a rogue DHCP or Radius server on the part of the attacker.

- This attack vector was recently engaged via "Trojan.AndroidOS.Switcher" Android malware which attacks wireless access points (36). When an infected device discovers its presence on a wifi network, it attempts to login to the wireless router administrative interface via user ID and password guessing. Once the malware accesses the wireless router, it modifies the DNS server parameters to point all the wifi network's DHCP clients to the attacker's DNS servers.

- Device infiltration – An attack to gain access to a device could provide the ability to edit the resolver configuration among other host information including installation of a root certificate signed by a rogue certificate authority (CA) operated by the attacker. This infiltration could be in the form of a brute force attempt to access the host resolver software or via malware.

Resolver redirection can be an effective means of resolving queries, to financial institutions, for example, to the attacker's site, which may proxy the session to the legitimate site to capture user information in transit as illustrated in Figure 4.8. The installation of the attacker's certificate enables seemingly legitimate secure (SSL) connection to the intended web server.

Individual client resolver attacks can severely impact the corresponding client device but generally have little impact on other DNS components as indicated in Table 4.6.

BROADER ATTACKS THAT LEVERAGE DNS

While several attack types target the DNS infrastructure itself, several broader network attacks leverage the DNS to inflict damage on other network components or to exfiltrate sensitive information outside the network.

TABLE 4.6   Client Resolver Attacks

Attack Type	Target Component	Potential Impact
Client resolver attacks	Stub resolver	Manipulation of the resolver configuration could take the resolver out of service or redirect queries to an attacker DNS server for this particular DNS client
	Forwarding DNS server	A client resolver attack should not impact forwarding DNS servers
	Recursive DNS server	A client resolver attack should not impact recursive DNS servers
	Referral DNS server	A client resolver attack should not impact referral DNS servers
	Authoritative DNS server for internal namespace	A client resolver attack should not impact authoritative DNS servers
	Authoritative DNS server for external namespace	A client resolver attack should not impact authoritative DNS servers

Network Reconnaissance

DNS by design houses a repository of hostname-to-IP address mapping among other things. If an attacker desired to glean information about particular hosts that may be more attractive to attack than others, he/she may start with DNS. Hosts named for the application or type of information contained therein (e.g., "payroll.example.com") may prove a desirable target. The following methods may be utilized by attackers to reconnoiter your namespace.

- Query sniffing – An attacker with access to the communications path to and from a given DNS server may log queries and answers in an attempt to identify potential targets.
- Name guessing – One brute force approach to such reconnaissance consists of guessing hostnames of interest and issuing standard DNS queries to obtain corresponding IP addresses if they exist.
- Wildcard (ANY) queries – An attacker may issue a query to your DNS server setting the QTYPE to "*" which is referred to as an ANY query. Servers configured to support this query, which most are by default, will typically respond with all of the resource records associated with the corresponding domain name (QNAME). This form of query may also be used in reflector attacks as discussed in the ensuing section.

- Zone transfers – Impersonating a DNS slave server and attempting to perform a zone transfer from a master is a form of attack that attempts to map or footprint the zone. That is, by identifying host to IP address mappings, as well as other resource records, the attacker attempts to identify targets for direct attacks.
- Next secure queries – If a given zone is signed via DNSSEC with the use of Next Secure (NSEC) resource records to support authenticated denial of resource record existence instead of the hashed NSEC3 version, an attacker may be able to identify hostnames in a zone by successively querying the zone for NSEC records to enumerate domain names. Potential impacts of attempts to perform network reconnaissance on different DNS components are summarized in Table 4.7.

TABLE 4.7   Network Reconnaissance Risks

Attack Type	Target Component	Potential Impact
Network reconnaissance	Stub resolver	Should have little impact as the resolver is not authoritative for any zone information, but if that of a "power user" certain queries and answers could prove useful
	Forwarding DNS server	Should have minor impact if an attacker monitors query and answer traffic to potentially identify targets opportunistically, but not a very efficient means of reconnaissance
	Recursive DNS server	Should have minor impact if an attacker monitors query and answer traffic to potentially identify targets opportunistically, but not a very efficient means of reconnaissance
	Referral DNS server	Should have no impact as referral servers offer up only general direction to a target domain but not specific host resolution data
	Authoritative DNS server for internal namespace	An attacker could request zone transfers, issue random or ANY queries to obtain resolution data; zones signed using NSEC could enable zone hopping to each resource record to identify potential targets
	Authoritative DNS server for external namespace	An attacker could request zone transfers, issue random or ANY queries to obtain resolution data; zones signed using NSEC could enable zone hopping to each resource record to identify potential targets

DNS Rebinding Attack

A DNS rebinding attack is so called because the resolution data for the same question is modified in the following manner. When a user browses to an attacker website, enticed by content, a phishing attack, social engineering, or other form of subtle coercion, the IP address resolved for the web address is the "legitimate" attacker web server IP address. The TTL for this RRSet is configured to a very short time interval. The corresponding web page contains malicious browser-side code such as flash or javascript.

When the browser code is executed, the code contains links to the website URL once again, which given the short TTL or the initial query, the resolver has already timed out of cache. Upon issuing a subsequent query to the attacker's DNS server as initiated by the browser-side code, the DNS server returns the IP address of an internal target, likely a private IP address. Thus, the IP address for the same domain name has been changed or rebound, to which the browser code launches its attack, possibly in the form of a DDoS or other attack.

This attack typically requires initial network reconnaissance using DNS or other form of discovery to identify the attack target. The target's IP address is used as the resolution RData for the attacker's web server during the rebinding phase. The querying of the attacker's domain repeatedly helps pass browser origin enforcement. Table 4.8 highlights the impacts of DNS rebinding attacks.

TABLE 4.8   DNS Rebinding Attack

Attack Type	Target Component	Potential Impact
DNS rebinding	Stub resolver	While the resolver is not typically the target of this attack, it performs its role of resolving attacker resolution data
	Forwarding DNS server	A DNS rebinding attack will update the cache entry for the attacker's IP address, which during the attack translates to an internal target typically
	Recursive DNS server	A DNS rebinding attack will update the cache entry for the attacker's IP address, which during the attack translates to an internal target typically
	Referral DNS server	A DNS rebinding attack should not impact referral DNS servers
	Authoritative DNS server for internal namespace	A DNS rebinding attack should not impact authoritative DNS servers
	Authoritative DNS server for external namespace	A DNS rebinding attack should not impact authoritative DNS servers

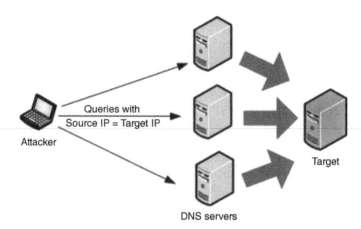

Figure 4.9.   Reflector Style Attack

Reflector Style Attacks

The reflector form of attack attempts to use one or more DNS servers to send massive amounts of data at a particular target as illustrated in Figure 4.9, thereby denying service for the target machine. Accomplishing this type of attack relies on leveraging DNS servers (or routes to DNS servers) which do not perform ingress IP filtering and on DNS servers configured to enable recursion. Typically, this form of attack features an attacker querying "open resolvers" or Internet-facing DNS servers configured to enable query recursion.

While recursion should be disabled for authoritative external DNS servers, unfortunately, there are millions of so-configured servers operating on the Internet today according to the Open Resolver Project (37). Upon receiving a query from a given IP address, each server will perform its recursion function and respond accordingly to the purported requesting IP address. This attack is analogous to a Smurf attack originated in the 1990s whereby an attacker would spoof the target IP address within ICMP (Internet Control Message Protocol used for "ping" and similar utilities) packets directed at numerous Internet servers to inundate the target with the spoofed IP address with ping responses.

- Reflector attack – The attacker issues numerous queries to one or more DNS servers using the target machine's IP address as the source IP address in each DNS query. This attack could be issued using authoritative or recursive DNS servers which will happily respond accordingly to the source IP address. If several servers are queried at the same time, the volume of DNS response packets can become very large.
- Amplification – Using the reflector approach while querying for resource record types with large quantities of data such as ANY queries, NAPTR, and DNSSEC-signed answers amplifies this attack by providing much larger

TABLE 4.9   Reflector Style Attacks

Attack Type	Target Component	Potential Impact
Reflector or amplification attacks	Stub resolver	Should have no impact on a reflector attack
	Forwarding DNS server	A reflector style attack should not impact forwarding DNS servers as it should never receive queries from the Internet
	Recursive DNS server	A reflector style attack should not impact recursive DNS servers as it should never receive queries from the Internet
	Referral DNS server	A reflector style attack should not impact referral DNS servers
	Authoritative DNS server for internal namespace	Unless the attack is directed at an internal target, this attack should have little impact as it should never receive queries from the Internet if properly configured
	Authoritative DNS server for external namespace	An external authoritative DNS server could be an unwitting participant in such an attack.

response packets. Each responding server responds with the data to the "requestor" at the spoofed IP address to inundate this target with a large data flow, amplifying the attack volume with respect to typical query answers (38). Potential impacts of reflector and amplification attacks upon the various DNS components are summarized in Table 4.9.

Data Exfiltration

Data exfiltration refers to the transmission of data originating from within one security domain, for example, an enterprise network, to another entity or organization, that is, the attacker's server. There are two basic forms of data exfiltration using DNS.

- The use of DNS as a data protocol to communicate between two endpoints through firewalls
- The use of DNS to locate external resources to which to convey information or obtain instructions for attack

DNS as Data Transport (Tunneling)   DNS tunneling entails the use of the DNS protocol as a data communications channel. This approach leverages the fact that DNS traffic is generally permitted through firewalls. This technique enables a user or device within the network to communicate with an external destination, easily traversing any intervening firewalls. Initially developed as a means to enable devices

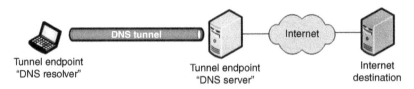

Figure 4.10.   DNS Tunneling

to communicate via a wifi network for free, the technology has also been used to exfiltrate information from enterprise networks by malware.

Utilizing the DNS protocol to tunnel data packets entails the client tunnel endpoint behaving as a resolver to issue a query to an "authoritative DNS server." The Qname section of the query contains the question in the form of end user data encoded as a domain name. The domain suffix corresponds to the domain for which the server-side tunnel endpoint is "authoritative." Let's say an attacker sets up a server endpoint using the domain: tunnel-example.net When registering the domain, the NS and glue records supplied to the parent domain (.net in this case) consist of the tunnel endpoint servers themselves.

A query to ns1.tunnel-endpoint.net, for example, will be directed to the tunnel endpoint server, as will link-to-my-email-or-website.tunnel-endpoint.net, a facetious expression of how client data would be transmitted to the server. The recursive server, root, and TLD servers, not shown in Figure 4.10, process the DNS query like any other, locating the name servers authoritative for tunnel-endpoint.net and directing the "query" to the server. The Qtype is usually TXT though NULL has also been used, though this RRType has been deprecated. Even an A or AAAA type can be used with return data in the form of a CNAME.

The tunnel server decodes the Qname and passes on the application request to the intended destination. Some level of fragmentation and reassembly may be required over the tunnel given DNS label length restrictions (63 octets) and full name length (255 octets). The client may post several A record queries to get the full file across, while the DNS server tunnel endpoint may reply with CNAME records with response data encoded within the RData portion.

Use of EDNS0 enables much longer responses, perhaps obviating the need for fragmentation and reassembly based on MTU support of the intervening infrastructure. The tunneling server processes responses from the Internet destination and encodes the response for transmission to the resolver. Typically, base32 encoding is used on outbound tunnel transactions and base64 in responses to support case sensitivity in the response octets, which the client tunnel endpoint decodes and presents back to the application software.

DNS as Resource Locator   An attacker may attempt to install malware on devices to enlist such devices under the control of the attacker as a botnet. Such malware may be installed via phishing or spear phishing attacks that bait users into

TABLE 4.10   Data Exfiltration Impacts

Attack Type	Target Component	Potential Impact
Data exfiltration attacks	Stub resolver	A device equipped with DNS tunneling software can participate in exfiltrating data via DNS
	Forwarding DNS server	Forwarding servers may participate in forwarding tunneling "queries" and responses
	Recursive DNS server	Recursive servers may participate in "resolving" tunneling queries and responses
	Referral DNS server	Data exfiltration attacks should not impact referral DNS servers
	Authoritative DNS server for internal namespace	Given the general objective to exfiltrate data, an internal authoritative server should not be impacted as it should have no external communications
	Authoritative DNS server for external namespace	Most attacks will feature an attacker-configured DNS server to serve as the tunnel endpoint as opposed to a hacked external authoritative DNS server with DNS tunneling software installed, so the impact should be minimal

opening executable email attachments or installing software from an attacker website. Whether a device is attacked while inside the enterprise network or a user device is physically brought onto the network, if it is trusted within the confines of an enterprise network it may have access to sensitive information. The malware may perform data collection, locating internal resources using DNS reconnaissance. In addition, DNS could be used to identify the current IP address of the attacker's external destination for exfiltration of the information.

As you can see from Table 4.10, the impact on DNS components themselves is minimal; however, the real threat is the theft of potentially valuable information.

Advanced Persistent Threats

Advanced Persistent Threats (APTs) are organized, stealthy forms of network intrusion where an attacker attains access within a target network to steal data, disrupt communications, or otherwise infiltrate network components. APTs are persistent in that the intent is to retain access to the network for a lengthy time frame, if not indefinitely, so they require continual evasion techniques to avoid detection.

Attackers may deploy malware within a network via phishing style attacks, social engineering, brute force hacking, or other methods. Malware installation on a device often goes undetected by the device's user. Simply opening an email attachment, downloading seemingly legitimate software, or clicking an advertisement may initiate the installation process. Virus protection software installed on your devices can help prevent such installation for known malware. But new forms of malware are continually developed to evade such protections and to install on the victim machine.

The attacker who successfully infiltrates a number of devices with malware can instigate such malicious code to perform operations on behalf of the attacker. These infected devices essentially serve as bots under the control of an attacker. A collection of such bots forming a botnet enables an attacker to utilize numerous devices potentially installed around the world to perform software programmable actions.

While attempting to avoid detection through stealthy activities, the attacker's bots typically need to communicate to the attacker's "command and control" (C&C) center. The C&C center is typically a server to which each bot connects to receive updates and commands to instigate attacks, update malware code, or collect information from the network in which a given bot resides. Typically, this process involves DNS queries to identify the IP address of the C&C center since it needs to be an Internet-accessible server for bot access.

If an IT administrator identifies the presence of malware within his or her network and can discern the C&C IP address based on DNS queries, he or she may block the IP address via a network or DNS firewall. To avoid such "easy" detection, many botnet administrators leverage the power of DNS to modulate the IP address and domain name of the C&C center to facilitate stealth. DNS component impact of APT attacks is summarized in Table 4.11 and we'll discuss the various methods of avoiding detection including DNS fluxing and dynamic domain generation in Chapter 11.

TABLE 4.11   Advanced Persistent Threats

Attack Type	Target Component	Potential Impact
APT attacks	Stub resolver	A device infiltrated with malware can participate in a botnet to perform nefarious functions
	Forwarding DNS server	Forwarding servers may participate in forwarding C&C queries and responses
	Recursive DNS server	Recursive servers may participate in resolving queries to C&C centers
	Referral DNS server	APT attacks should not impact referral DNS servers directly though they may use referral services
	Authoritative DNS server for internal namespace	APT attacks should not impact internal authoritative DNS servers directly though they may use DNS services to locate potential targets (reconnaissance)
	Authoritative DNS server for external namespace	APT attacks should not impact external authoritative DNS servers

SUMMARY

As we've seen, several varieties of DNS attacks are possible to disrupt DNS or net-work communications in general or to leverage DNS' intended purpose to identify targets or attack systems with malicious intent. No single mitigation approach can eliminate vulnerabilities to all threats; thus, a multipronged mitigation strategy is required to reduce attack exposure. The ensuing chapters drill down into more detail on each of these threat areas and discuss tactics you may employ to help defend your network by defending your DNS.

5

DNS TRUST SECTORS

INTRODUCTION

The approach we apply in this book to securing DNS mimics and attempts to comply with broader information technology (IT) network security approaches. Two basic tenants of such an approach entail partitioning DNS server deployments and corresponding functions based on *trust zones* and employing a multilayered *defense in depth* style approach. We'll use the term "trust sectors" instead of "trust zones" given the ambiguity of the word "zone" in a DNS context. Establishing an effective defense is critical as is preparation, monitoring, event detection to rapidly identify attacks in progress, and enacting recovery plans to perform mitigation actions to minimize or nullify their impacts. Event postmortems are also critical to feeding back to the security plan to apply lessons learned to improve detection and recovery times.

This chapter focuses on the identification of trust sectors within your DNS infrastructure to enable resolution of your users' queries for internal or external destinations, and to enable resolution of your external namespace by global Internet users. Implementing trust sectors implies deploying DNS servers within each sector to perform particular functions. Deploying DNS servers in such a manner helps contain security breaches to the given sector, minimizing impacts on other sectors. We'll describe deployment strategies that effectively partition DNS information and

DNS Security Management, First Edition. Michael Dooley and Timothy Rooney.
© 2017 by The Institute of Electrical and Electronic Engineers, Inc. Published 2017 by John Wiley & Sons, Inc.

communications in order to contain vulnerabilities and attacks within these respective trust sectors. We'll discuss basic firewall policy settings that serve as a starting point to partition these trust sectors from a networking perspective and DNS filtering tactics at the DNS protocol level that are provided by various DNS server software products.

Generally, DNS deployment designs should account for high availability, performance, scalability, human intervention, and of course, security. Using a trust sector approach to DNS server deployment allows you to segment namespace and resolution responsibility which provides a solid foundation for achieving these objectives. Keep in mind that there is no "one size fits all" cookie-cutter deployment architecture. However, by defining role-based server configurations as trust sectors, you can select which are applicable based on your environment's scale and policies.

Some general deployment principles to keep in mind include the following:

- Deploy a master DNS server and at least two slaves as authoritative for any given zone or set of zones. For multi-master replicated DNS server implementations, deploy multiple masters for each set of zones.
- For implementations not natively supporting a multi-master approach, consider deploying redundant hardware for the master to minimize impacts of a master server outage. Should a master server become unavailable, slaves may still effectively resolve queries authoritatively at least until the zone expires (as defined in the zone's SOA record). An alternative master could be promoted in order to effect any zone changes with replication to slave (secondary) servers.
- Deploy servers that are authoritative for a set of zones each on different subnets and ideally, different locations for site-diverse high availability. Should a subnet or router become unreachable, DNS services should be available from alternative sites.
- Deploy authoritative servers "close" to clients/resolvers for better performance and less network overhead. For external servers, deploy close to Internet connections; for internal servers, deploy nearer to higher density employee areas.
- Consider anycast deployment to provide redundancy as well as potentially improved resolution performance. Consider load balancing deployment as well to optimize performance.
- To provide functional separation, different DNS servers should be deployed to handle external queries versus internal queries and for handling recursive versus authoritative queries. This principle is critical to deployment of DNS trust sectors, to maintain functional separation and granular access controls.
- Deploy dedicated recursive servers to support client/stub resolver resolution. You may want to consider a tiered recursion model as we'll discuss later in this chapter.

CYBERSECURITY FRAMEWORK ITEMS

We'll begin to apply the NIST cybersecurity framework in this chapter as a baseline, which we will frame in the context of DNS. A complete example model framework profile can be found in Appendix A for your reference. Successive chapters will add descriptions of relevant framework items. The following sections summarize the NIST cybersecurity framework categories that we apply to DNS trust sectors.

Identify

Physical Devices and Systems are Inventoried (ID.AM-1)   All DNS components including those systems housing resolver software and DNS servers themselves must be documented and tracked. Tracking of device-specific identifiers such as serial numbers along with installation location down to the rack location if applicable should be performed and maintained.

Software Platforms and Applications are Inventoried (ID.AM-2) Operating system and DNS software running on your physical DNS servers must be inventoried with respect to vendor, release number, patches applied, and last verification date.

Organizational Communication and Data Flows Are Mapped (ID.AM-3)   You must document your DNS server deployment with respect to itemizing and inventorying each server, its role and the communication and data flows for DNS protocol and DNS management transactions. We'll discuss these flows in this chapter.

External Information Systems Are Cataloged (ID.AM-4)   If you use a service provider for your external authoritative DNS or for DNS recursion, identify and track relevant service provider information including SLAs and contact information as well as technical details regarding IP addresses and security features.

Protect

Network Integrity Protection with Network Segmentation (PR.AC-5) A key tenant of defining a trust sector deployment approach entails segmentation of DNS servers by function to simplify configuration and management, and to help contain any intrusions.

Adequate Capacity to Ensure Availability (PR.DS-4)   One key aspect of deployment planning entails provisioning of DNS capacity for recursion or resolution while accounting for outage risks due to attacks, natural disasters, or human error.

Communications and Control Networks Are Protected (PR.PT-4)
Trust sector deployment attempts to protect sets of DNS components by function and
we outline this approach along with associated network firewall support functions in
this chapter.

Detect

Baseline Network Operations and Expected Data Flows (DE.AE-1)
Deployment plans seek to provide adequate DNS resources for acceptable DNS resolution performance while accounting for potential outage conditions. Ongoing monitoring of DNS protocol and DNS management data flows enables tracking of actual
demand with respect to provided capacity and provides feedback to justify supple-
mentation of relevant capacity and resources.

DNS TRUST SECTORS

We define four major trust sectors based on

a. Query source: from where a query originates
b. Query scope: the scope of information being queried

We define the query source as either external queries originating from outside your
organization, for example, the public Internet, which generally has low to no trustworthiness, or internal queries originating from within your organization, which may
possess moderate trust. The query scope also follows an analogous breakdown, with
external scope dealing with Internet-reachable resolution data and internal encom-
passing resolution information for destinations within your organization. Figure 5.1
summarizes this categorization.

You may also overlay a DNS management trust sector which may correspond to
a physically or virtually separate network used to configure and manage each DNS
server. In addition, we'll discuss a few outlier scenarios that apply across all trust

		Query scope	
		External	Internal
Query source	External	**External DNS**	**Extranet DNS**
	Internal	**Recursive DNS**	**Internal DNS**

Figure 5.1.   Basic DNS Trust Sectors

sectors such as the aforementioned DNS management sector. These may apply to one or more categories as they provide special resolution or availability features.

A summary of each of these trust sectors is as follows.

- **External DNS Sector**– This category consists of DNS servers deployed to resolve queries originating from the Internet for your public resolution information, that is, your external namespace. If you have an Internet connection for a website, email, or for other publicly available Internet applications, this category must be addressed in your deployment strategy.
  - **External DNS Server Deployment** – The deployment scenario corresponding to this sector seeks to provide robust name resolution for external clients seeking legitimate name resolutions for the organization's public resources, such as web servers, email servers, and the like, while minimizing exposure to those seeking to attack the DNS infrastructure or infiltrate it for attack purposes.

    Deployment of external DNS servers features a hidden master server with a number of slave servers. As we'll see, these servers should never be queried by a resolver directly; only by recursive name servers resolving on behalf of resolvers. As such external DNS servers should never be configured to support recursion. These servers are exposed directly or thinly veiled within a DMZ and are therefore susceptible to attacks of most types.

    As an alternative to in-house deployment and management of external DNS servers, some organizations opt to use an external DNS service provider, which provides a web user interface for the management of externally resolvable zones and resource records.
- **Extranet DNS Sector** – This sector includes queries from specific sources outside the organization seeking resolution for internal hosts and resources. Such resolution should generally be forbidden, but organizational partners may require secure access to certain servers that aren't available publicly. With the provision of such partners' access to a subset of "internal" resolution information, this sector is marginally of higher trust than the external sector. DNS server deployment for this category (for partner access) should mimic the external DNS scenario, though possibly deployed as a parallel per-partner implementation.
  - **Extranet DNS server deployment** – Partners presumably access your network via virtual private network (VPN) and you can publish IP-reachable resource availability via extranet DNS servers deployed in your partner VPN DMZ. These servers should be secured equivalently to external DNS servers as attackers may attempt to access your VPN (partner credential security is out of your direct control). Resource records published within these DNS servers should enumerate only resources to which you desire to provide access and name resolution.

- **Recursive DNS Sector** – This sector consists of queries originating from within your organization requesting external/Internet resource resolution.
  - ○ **Recursive DNS Server Deployment** – Recursive or caching servers are internal DNS servers that cache resolutions for use by internal DNS servers on behalf of internal resolvers. Caching servers should be deployed independently of internal authoritative DNS servers. Queries for internal namespace should be directed and/or forwarded to the internal authoritative servers, while all other queries should leverage cache or the servers' hints files to query Internet root servers down the domain tree to resolve queries, building up a cache of resolved data.

    Alternatively, distributed recursive servers deployed throughout the organization can be configured to forward internal namespace queries to internal authoritative servers and all other queries to a second tier of recursive servers. This second tier of servers enables other internal-resolving DNS servers to be configured to funnel queries for external data through these servers. Doing so enables more control over which servers are permitted to issue external queries while enabling them to build up a substantial cache over time.
- **Internal DNS Sector** – This sector deals with internally originating queries for internal resolution information.
  - ○ **Internal Authoritative DNS Servers** – DNS servers are required to resolve queries for internal destinations from internal hosts. These DNS servers are configured with authoritative information for the internal namespace. As with external master DNS servers, internal master DNS servers should be "hidden" for added security and information integrity.
  - ○ **Departmental Authoritative DNS Servers** – For larger organizations, some business units or entities may desire to run their own name subspace within the organization's namespace. This scenario features delegation of namespace internally but is otherwise a replica of the internal resolution DNS servers case, though for a subset of the internal namespace. In this sense, this is merely a special case of internal authoritative DNS servers with an internally delegated zone or set of zones.
  - ○ **Internal Root Servers** – Internal root servers can be configured as the authoritative root of the internal namespace for resolution of internal queries. The intent of an internal root is to eliminate queries from internal sources from reaching the Internet root servers or Internet DNS servers in general. This type of deployment helps secure DNS resolution by constraining all internal queries within the internal network, but only enables resolution of domain names provisioned within subdomains of the internal root.

Based on the size of your organization (and budget), you should deploy an external set of DNS servers (or use an external DNS provider), a set of internal authoritative servers, and a set of recursive servers. This is the minimum trust sector deployment configuration. Larger or more sophisticated deployments may feature tiered approaches and possibly extranet DNS. Let's explore each of these sectors in more detail in the context of network deployment.

EXTERNAL DNS TRUST SECTOR

The external DNS trust sector relates to servicing DNS queries originating outside or external to the organization. Resolution services must be provided for your external namespace, that is, your organization's website, email, and other applications. But care must be taken to secure the information integrity of these external servers, given their inherent exposure and potential vulnerability in serving external clients.

The recommended approach is to deploy two or more slave DNS servers to resolve external requests, and to configure these servers with IPv4 and IPv6 addresses. These slave servers may be deployed directly on an external DMZ exposed to the Internet, behind a "first line" firewall, as shown in Figure 5.2. Note that the inside and outside firewalls depicted in Figure 5.2 may physically be a single firewall device, but we'll use this logical view for clarity. If you have multiple ISP access links, you should deploy at least one slave DNS server at each DMZ.

In this example, we've configured each DNS server as dual-stack (IPv4 and IPv6) to enable reachability via either protocol. We've placed each external DNS server on its own physical subnet and ideally router/interface with public IP addresses as shown, though internally reachable IP addresses must also be assigned for server management and zone transfers from the master.

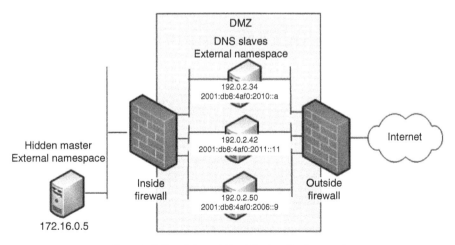

Figure 5.2. External Trust Sector Deployment

Figure 5.2 illustrates a *hidden master* DNS server deployed behind a DMZ internal firewall and should not be directly query-able from external clients. Since this master server maintains the "master configuration" from which the slave servers transfer, its information integrity must be safeguarded. For this reason, this master DNS server should be configured as hidden, meaning that it cannot be identified by querying other DNS servers.

Hiding the master DNS server reduces the risk of an attacker identifying the master server, then attempting to infiltrate its configuration. Imagine the potential impact if an attacker changed your www record to an illicit website! The mechanics of hiding a master name server entail excluding NS and glue records for the hidden server in this and the parent domain zone file and modifying the master server name ("mname") field of the SOA record in each zone db file. The mname field typically enables an entity desiring to update the zone to locate the DNS server to which to direct the update. External-facing zones are generally static zones, disallowing dynamic updates, so modifying the mname field should have no repercussions.

This external trust sector should be isolated by setting appropriate firewall rules on both inside and outside firewalls. With respect to enabling DNS traffic, the outside firewall should be configured to enable incoming DNS queries from any source with a destination address matching any of the DMZ slave DNS servers on destination UDP and/or TCP port 53. Responses likewise need to be permitted. An example external firewall rule set is shown in Table 5.1. These DNS servers must be configured to disallow recursive queries.

As for the interior firewall, ACLs should be configured to deny all DNS queries originating from the Internet. This firewall should only permit DNS queries and answers between the hidden master and slave DNS servers for zone maintenance, including refresh queries, notify and zone transfer messages. Note that major DNS

TABLE 5.1   Example External Firewall Rules for DNS Messages

Message/ Direction	Control	Source Address	Source Port	Destination Address	Destination Port
DNS queries from the Internet	Allow	Any	>1023	192.0.2.34, 192.0.2.42, 192.0.2.50, 2001:db8:4af0:2010::a, 2001:db8:4af0:2011::11, 2001:db8:4af0:2006::9	53
Responses to DNS queries	Allow	192.0.2.34, 192.0.2.42, 192.0.2.50, 2001:db8:4af0:2010::a, 2001:db8:4af0:2011::11, 2001:db8:4af0:2006::9	53	Any	>1023
All others	Deny	Any	Any	Any	Any

TABLE 5.2   Example Internal Firewall Rules for DNS Messages

Message/ Direction	Control	Source Address	Source Port	Destination Address	Destination Port
Queries from slaves to master for zone maintenance	Allow	DMZ DNS server private IP addresses	>1023	172.16.0.5	53 or configured
Responses from master to slaves from zone maintenance	Allow	172.16.0.5	53 or configured	DMZ DNS server private IP addresses	>1023
All others	Deny	Any	Any	Any	Any

implementations enable you to define a specific port number if desired for notify and zone transfer messages. Table 5.2 provides an example interior firewall rule set. Additional firewall permissions need to be set for outbound queries for the recursive trust sector as discussed later.

Basic Server Configuration

Securing trust sectors from a networking perspective calls for the application of policies not only within relevant firewall configurations but also within the DNS server application you choose to deploy. The following features should be implemented to the extent your chosen product supports them:

- Disallow recursive queries
- Sign zone transfers between the master and slaves' private IP addresses
- Allow notify's and zone transfers only among the master and slaves' private IP addresses
- On the hidden master, allow queries only from the slaves' private IP addresses
- Prevent administrative access except from the "management" (i.e., internal) IP address space
- Inhibit exposure to the server implementation (version queries) to the extent possible by disabling vendor and version information; for example, via dig @<dns_server_ip_address> version.bind chaos txt.
- Secure the server (discussed in Chapter 6)
- Defend against distributed denial of service (DDoS) and reflector attacks (Chapter 7)

- Protect authoritative data from attack including signing zone information via DNSSEC (Chapter 9)

ISC BIND   The following is an example BIND configuration subset highlighting BIND-level controls for external DNS slave servers.

```
# sample configuration of slave 192.0.2.34
# define a TSIG key between this slave and the master

key "master-slave.ipamworldwide.com." {
    algorithm hmac-sha256;
    secret "MKWCLOu2s4IsZy60sH4q6a4qAud3xk0rinx4FVKut/0=";
};


# associate the TSIG key with the master to sign communications

server 172.16.0.5 {
    keys { "master-slave.ipamworldwide.com." ;};
};


# disable the rndc control channel

controls { };
# configuration options to disable recursive query processing,
# provide a null response to version queries, accept notify's
# only from the master, ignore requests for zone transfers and
# to perform zone transfers from the master on its internal
# IP address

options {
    recursion no;
    version "none";
    allow-notify { key "master-slave.ipamworldwide.com." };
    allow-transfer { none; };
    transfer-source 172.16.1.34;
};
zone "ipamworldwide.com."
    type "master";
    file "db.ipamworldwide.com.signed"
};
```

NLnet NSD   An NLnet NSD server example configuration subset follows below. Note that NSD is an authoritative only DNS server and cannot be configured to perform recursion. The ACLs need to be defined on a per zone basis though a common zone *pattern* can be defined and applied across several zones.

```
# sample NSD server configuration (version 4.1.11)
# server options below ignore version queries

server:
     hide-version: yes

# disable the control channel (which is the default)

remote-control:
     control-enable: no

# define a TSIG key

key:
     name: "master-slave.ipamworldwide.com."
     algorithm: hmac-sha256
     secret: "MKWCLOu2s4IsZy60sH4q6a4qAud3xk0rinx4FVKut/0="
# for each zone define its name and file and options including
# from where notify's are permitted (signed), to whom to issue
# zone transfer requests and on what outbound address to query
# for and receive zone transfers

zone:
     name: "ipamworldwide.com"
     zonefile: "db.ipamworldwide.com"
     allow-notify: 172.16.0.5 master-slave.ipamworldwide.com.
     request-xfr: 172.16.0.5 master-slave.ipamworldwide.com.
     outgoing-interface: 172.16.1.34
```

PowerDNS   The PowerDNS Authoritative server enables resolution of zone data that may be stored in any number of "backends" ranging from stock BIND zone files to relational databases like MySQL or Oracle among others. As such, zone-based parameters such as TSIG keys are stored in the backend. The relational database schema calls for configuring such information in domain metadata.

For example, to define the same TSIG key we've been using and require its use in requesting a zone transfer from a master DNS server, one could insert the following records into the backend via SQL:

```
$ insert into tsigkeys (name, algorithm, secret) values
('master-slave.ipamworldwide.com.', 'hmac-sha256',
'MKWCLOu2s4IsZy60sH4q6a4qAud3xk0rinx4FVKut/0=');

$ select id from domains where name='ipamworldwide.com';
7
```

```
$ insert into domainmetadata (domain_id, kind, content)
values (7, 'AXFR-MASTER-TSIG', ' master-slave.ipamworldwide.com.');
```

First the key is defined with the insert into the `tsigkeys` table. Then having identified the domain id for our ipamworldwide.com domain, we insert the key into the domain's metadata using the `AXFR-MASTER-TSIG` metadata type. The pdns.conf file enables configuration of database connectivity parameters to enable PowerDNS to connect to the backend. Some DNS settings can also be defined within the pdns.conf file or on the command line when invoking the pdns binary.

```
allow-notify-from=172.16.0.5
allow-recursion=none
version-string=anonymous
```

Knot DNS   A comparable example Knot DNS server configuration follows. Knot produces a separate authoritative only and recursive server. In this example, using the authoritative version, recursion is disabled by definition.

```
# sample Knot DNS server configuration (version 2.3.3)
# ignore version queries
server:
  version: none

# define a TSIG key
key:
  - id: master-slave.ipamworldwide.com.
    algorithm: hmac-sha256
    secret: "MKWCLOu2s4IsZy60sH4q6a4qAud3xk0rinx4FVKut/0="

#define a notify acl
acl:
  - id: accept_notify
    address: 172.16.0.5
    key: master-slave.ipamworldwide.com.
    action: notify

#sign all transactions with the master
remote:
  - id: master
    address: 172.16.0.5
    key: master-slave.ipamworldwide.com.

zone:
  - domain: ipamworldwide.com
    storage: /var/lib/knot/zones/
```

```
# file: ipamworldwide.com.zone    # Default value
master: master
acl: accept_notify
```

DNS Hosting of External Zones

Some organizations prefer to outsource configuration and management of their external namespace. DNS hosting providers typically offer site-diverse anycast-addressed DNS servers to host customer zone information. A third-party provider offers the convenience of offloading internal resources otherwise required for physical servers, expertise in configuring external DNS and in monitoring and managing external DNS servers and configuration.

You can also deploy a hybrid configuration where your hosting provider provides added redundancy by hosting secondary/slave servers to your in-house master(s) or conversely hosting masters and slaves, in conjunction with in-house secondary/slave servers. Be sure to follow similar guidelines as discussed above, as well as those controls for other vulnerabilities we'll discuss in subsequent chapters, when configuring your in-house DNS servers to securely interact with your hosting provider's DNS servers and the Internet at large.

When selecting an external DNS hosting provider, keep in mind the following security requirements:

- Unique per user login/password access
- Encrypted connection for administrator access
- Administrator access logs which can be reviewed and audited
- DNSSEC signing with planned and emergency key rollover support
- Other DNS security features including ACLs (i.e., no recursion, allow-transfer, etc.), geographic resolution, and response rate limiting (which we'll cover in later chapters)
- DNS availability support and service-level agreement (SLA)
- DNS denial of service mitigation
- Parent domain (typically TLD) security controls and vulnerability/breach notification process

External DNS Diversity

Supporting a web presence requires a robust infrastructure including external DNS. If your organization relies on the web for commerce, collaboration, or communications, deploying a robust external DNS infrastructure is paramount. We've mentioned the requirement to deploy multiple DNS servers dedicated to the external DNS trust sector. But you may want to consider adding further diversity to provide high availability and robustness in the face of attacks or other vulnerabilities such as human error or natural disasters. Consider implementing the following diverse components:

- Deploy multiple DNS servers deployed in different geographic locations
- Implement multiple DNS server vendor implementations to protect against attacks on a given vendor's vulnerabilities
- Use multiple external DNS providers or supplement your in-house implementation with an external service provider

While managing a diverse external DNS infrastructure may cost more and require incrementally higher management effort, this approach can help your external DNS trust sector withstand a variety of attacks.

EXTRANET DNS TRUST SECTOR

The extranet trust sector comprises external partner hosts querying information regarding partner-related (nonpublic) resolution information. In general, divulging information about internal hosts is undesirable and a potential security risk particularly within the realm of hostname reconnaissance. Even interconnected partners should only have access to guarded information, certainly not the entire internal namespace. Thus, the extranet trust sector is only incrementally less restrictive than the external trust sector.

Inter-partner connections are typically configured as virtual private network (VPN) connections over the Internet or private network and typically involve a "partner DMZ" or firewall between the partner space and the internal network, similar to the external design. Thus, the DNS deployment architecture for this category, shown in Figure 5.3, mimics that of the external DNS deployment though the resolution data configuration is somewhat differently. Depending on what resolution data may be divulged to a given partner, the DNS server queried by partner clients must be configured accordingly with such data. Only systems to which each partner is authorized access should be published in DNS. Thus, the concept of a hidden master with visible slaves supporting no recursion per the external sector applies.

The partner-specific resource record information may be defined within an "extranet" namespace, contained within respective zone files configured on these DNS servers. Additionally, implementing views on the DNS servers serving the partner link enables per-partner resolution information if multiple partners access a common set of DNS servers. DNS views allow the DNS server to answer queries depending on "who's asking" in the match-clients statement and "whom they're asking" with the match-destinations statement. In this manner, resolving a given hostname for a Partner A client may differ from that query of the same hostname from a Partner B client. One caveat relates to partners' use of common VPN termination hardware or

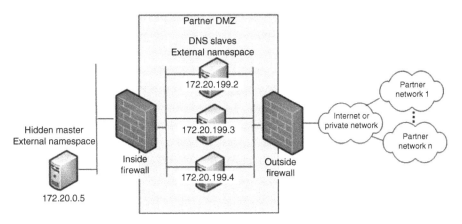

Figure 5.3. Extranet Trust Sector

address space, as the use of views requires the mapping of separate IP address space for each partner. DNS views are supported on ISC's BIND but not on NSD or PowerDNS authoritative products, though separate server instances may be run for each partner if desired.

Each partner's DNS configuration should be defined to forward to your extranet DNS servers to resolve information that you wish to divulge about your network. The extranet trust sector is isolated in a manner similar to the external trust sector. The outside firewall should be more restrictive though on sources of queries, limiting these to known partner querying DNS server addresses. The inside firewall should be configured from a DNS standpoint to enable DNS transactions only between the master and slaves. Additional policies are required to enable outbound queries to partner DNS servers and responses.

From a DNS server configuration, parameters specified in the external sector section apply with the addition of adding access control lists (ACLs) on the query source IP address space. If the partner systems querying your DNS servers reside within a common subnet or CIDR block of address space, you could specify the allow-query option or its equivalent, to scope from what partner addresses queries are permitted though if NAT'd via VPN connection, it may offer little defensive resistance.

RECURSIVE DNS TRUST SECTOR

This trust sector comprises internal stub resolvers querying internal DNS recursive servers, which in turn query internal and external referral and authoritative DNS servers to resolve client queries. Servers within this sector are not authoritative for any zone information and are configured solely to resolve queries on behalf of resolvers.

Intervening DNS forwarder servers may also be deployed, providing a tiered caching architecture.

Tiered Caching Servers

Deployment of DNS servers dedicated to recursion and caching is a recommended approach to provide functional, physical, and administrative separation from servers in other trust sectors. Caching servers deployed near client populations can facilitate rapid resolution performance once cache has been primed. However, on the flip side, if several such servers are deployed throughout your network, each issuing DNS queries through your firewalls to Internet DNS servers, the task of controlling and monitoring query traffic could quickly become cumbersome.

Internet name resolution requires IP (DNS) traffic outbound from the organization to the Internet, which may increase exposure from a security policy perspective. And many servers in different locations may issue redundant queries for the same resolution information, reducing efficiencies. One approach to alleviate these concerns entails the use of a set of tiered caching servers as a second layer through which all outbound queries can be issued. The first tier comprises local recursive servers to which local stub resolvers direct queries. These local recursive servers can be configured to forward all or certain queries to the second-tier servers, which consist of a set of what we'll call "Internet caching" servers. These Internet caching servers then query DNS servers on the Internet to resolve queries and cache answers. Similar queries from different local recursive servers can leverage the broader cache accumulated by the Internet caching tier without requiring Internet DNS lookup, thereby improving performance and reducing the volume of Internet DNS transactions.

Internet caching servers serve as funnel points to resolve queries from local recursive servers for information outside of the internal namespace. The deployment of Internet caching servers not only helps constrain the sources of outbound queries, but simplifies configuration of firewalls for Internet DNS queries from internal sources by reducing the number of valid querying IP address sources. Other name servers within the organization will forward queries to these caching name servers when they are unable to resolve directly from authoritative configuration or their own cache.

Internet caching servers should be deployed in a high availability configuration, due to the reliance on these servers for resolving Internet queries on behalf of internal hosts. Since these caching servers will frequently send and receive Internet traffic, they should be deployed close to Internet connections. Adding this to our previous external DNS figure, Figure 5.4 illustrates deployment of a high availability pair within the internal network but relatively close to the Internet connection. If you have two diverse Internet connections, as with external DNS servers, it's a good idea to deploy a server or pair near each connection.

While the external servers resolve queries for your public information for external queriers, the Internet caching servers resolve external information on behalf of your internal clients. The Internet caching name servers' public IP addresses need to

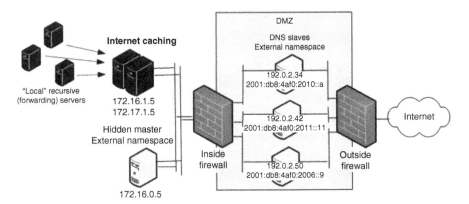

Figure 5.4.   Addition of Caching Servers for External Resolution

be added to the firewall permit lists to enable resolution of Internet host names for internal clients as illustrated in Table 5.3. The use of one or a small number of such name servers enables specification of only these few addresses instead of every DNS server address within the organization that would otherwise execute iterative queries. This policy to permit outbound queries and responses from/to Internet caching servers needs to be applied to both inside and outside firewalls. Configuring IP address filtering using reverse path forwarding on your (and hopefully your ISP's) routers in accordance with BCP38 (39) can help reduce the success of spoofing.

Basic Server Configuration

Caching server configuration for DNS processing is relatively trivial. There are no zones or resource records to configure (other than the root hints and localhost-related zones). But configuration for security purposes and monitoring of query activity over time are critical to effectively managing DNS security. We'll cover the basics

TABLE 5.3   Example External Firewall Configuration for DNS Caching Messages

Message/ Direction	Control	Source Address	Source Port	Destination Address	Destination Port
Internet caching server queries	Allow	NAT'd (172.16.1.5)	>1023	Any	53
Responses to Internet caching server queries	Allow	Any	53	NAT'd (172.16.1.5)	>1023
All others	Deny	Any	Any	Any	Any

here, then delve more deeply in respective chapters. The following settings should be configured:

- Allow recursive queries only from lower tier forwarder DNS servers (local recursive servers) and/or internal clients using your allocated internal (e.g., private) address space.
- Allow query access to cache to lower tier forwarders and/or internal clients.
- Allow recursion, queries, and access to cache only on the server interface possessing the internal IP address. This will help prevent spoofed queries received on other server interfaces (e.g., DMZ facing).
- Prevent externally spoofed query packets by configuring router/firewall IP address filtering using reverse path forwarding.
- Disallow dynamic updates and zone transfers.
- Prevent administrative access except from the "management" (i.e., internal) IP address space.
- Inhibit exposure to the server implementation to the extent possible by disabling vendor and version information; for example, via dig @<dns_server_ip_address> version.bind chaos txt.
- Secure the server (discussed in Chapter 6).
- Define query rate limits (Chapter 7).
- Configure DNSSEC validation (Chapter 8).
- Configure DNS firewall (Chapter 11).

Sample BIND Configuration   Here's an example simple BIND configuration.

```
acl internal-nets { 10.0.0.0/8; 172.16.0.0/12; } ;
options {
      recursion yes;
      version "hidden";
      allow-recursion { "internal-nets"; };
      allow-recursion-on { 172.16.1.5; };
      allow-query { "internal-nets"; };
      allow-query-cache { "internal-nets"; };
      allow-query-on { 172.16.1.5 };
      allow-transfer { none; };
      allow-update { none; };
};
zone "." {
      type hint;
      file "root-hints.file";
};
```

Only the root zone is shown in this named.conf file snippet above, and other than localhost zones (localhost, 0.0.127.in-addr.arpa and 0.0.0.0.0.0.0.0.0.0.0.0.0.0. 0.0.0.ip6.arpa), no other authoritative zones should be configured.

Sample Unbound Configuration   Unbound is the NLnet recursive server and its analogous configuration might look like this.

```
server:
    hide-version: yes
    interface: 172.16.1.5
    access-control: 10.0.0.0/8 allow
    access-control: 172.16.0.0/12 allow
    root-hints: "root-hints.file"
```

The access-control statement defines an ACL regarding from what IP addresses queries to the cache will be processed.

PowerDNS Recursor Configuration   Entries in the pdns.conf file such as the following enable query access control, return "hidden" to version queries, and defines the root hints file.

```
allow-from=10.0.0.0/8, 172.16.0.0/12
version-string=hidden
hint-file=root-hints.file
```

Knot DNS Resolver Configuration   Knot DNS Resolver configuration syntax is written in the Lua language which enables you to define conditional operations if desired. The following relevant configuration parameters can be set as shown using the command line after running kresd.

```
net.listen('172.16.1.5')
modules = { 'hints', 'policy' }
hints.config({file='/etc/root-hints.file')}
policy.add(policy.all(policy.PASS, '10.0.0.0/8'))
policy.add(policy.all(policy.PASS, '172.16.0.0/12'))
```

INTERNAL AUTHORITATIVE DNS SERVERS

DNS servers must be deployed to resolve queries from internal clients for internal host information. We can refer to the zones they serve as internal second-level domains (SLDs) since they will be the masters of the second level (just below TLD) of the internal namespace, for example, ipamworldwide.com., and may delegate subdomains to other internal DNS administrators and respective DNS servers. As with the external trust sector, the internal master DNS server should be deployed as a hidden master. This will improve the information integrity of master servers by hiding them, as internally initiated attacks account for a moderate percentage of network security breaches.

Deploying a sufficient number of slave servers, authoritative for respective internal zone information, enables resolution of client queries, while offloading the master servers to handle only configuration updates. If a master DNS server fails, slaves will continue to resolve queries, but a lengthy outage can compromise the validity and certainly timeliness of the slaves' zone data. The slaves will continue supporting this zone data until the zone expiration time is reached, after which the server will no longer consider itself authoritative for the zone. Dynamic updates will also not be possible if the master server is down.

An important consideration for Microsoft client environments with client-driven dynamic updates when attempting to hide a master DNS server is that Microsoft clients rely on the master DNS server name (mname) field of the zone's start of authority (SOA) resource record to identify the master DNS server to which to send an update. In this case, you can still hide the master by changing the mname field to point to a legitimate slave server if the slave server supports update forwarding, to forward updates to the primary master. In general, we recommend against having clients directly update DNS in favor of having your DHCP servers perform this function when assigning IP addresses to devices. The fewer the entities that can update DNS, the tighter the access security can be configured and the fewer the variety of update sources will be able to impact DNS data integrity.

The use of DHCP servers is generally necessary anyway to provide dynamic addressing to laptops, desktops, printers, mobiles, VoIP phones, and other IP devices. Given that most, if not all, of these device types will require entries in DNS corresponding to their respective assigned addresses, we need to allow updating of DNS from our DHCP servers. Since we have a hidden master, we could configure the DHCP servers to update a slave DNS server. This server can be deployed with hardware redundancy to minimize any outage time where DNS cannot be updated for DHCP clients.

Figure 5.5 shows an example four-server deployment for our internal ipamworldwide.com namespace. As described in the architecture overview, internal client resolvers should be configured with at least two DNS servers. Any number of additional slaves can be deployed in branch offices or remote sites to balance the query load. In general, we do not specify intervening firewalls in this architecture given all servers reside internal to the enterprise network, but you may deploy them for added protection. In any case, none of these servers should accept DNS queries from outside the organization's address space.

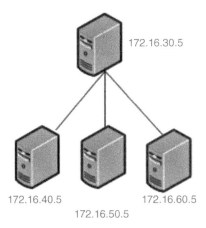

172.16.30.5

172.16.40.5

172.16.50.5

172.16.60.5

ipamworldwide.com zone

Figure 5.5. Internal DNS Servers for Internal Clients

Basic Server Configuration

Internal authoritative server configuration should mimic the external server configuration to provide adequate security controls and also consistency, which can help reduce administrator errors. The following features should be implemented to the extent your chosen product supports them:

- Disallow recursive queries
- Sign zone transfers between the master and slave servers
- Allow notify's and zone transfers only among the master and slaves
- For the hidden master, allow queries only from the slaves' IP addresses
- Prevent administrative access except from the "management" IP address space
- Inhibit exposure to the server implementation to the extent possible, for example, disable vendor and version information, for example, via dig @<dns_server_ip_address> version.bind chaos txt
- Secure the server (discussed in Chapter 6)
- Defend against DDoS attacks (Chapter 7)
- Protect authoritative data from attack and consider signing with DNSSEC (Chapter 9)

ISC BIND Example The following is an example BIND configuration subset highlighting BIND-level controls for internal DNS slave servers. In this example, we'll enable the control channel but secure it with an ACL and rndc key. We'll

assume our management network from which rndc commands are authorized resides
on 172.20.254.0/24, and define our ACL "mgmt.-net" accordingly. Here, we'll use a
separate server interface, listening on 172.16.50.6, as the control channel.

```
# sample configuration of slave 172.16.50.5
# define and address match list and key for control channel
acl mgmt-net { 172.20.254.0/24; };
key "internal-2-rndc.ipamworldwide.com." {
        algorithm hmac-md5;
        secret "UJfGhXM1e1YFLC5UJI0MAQ==";
};

# define a TSIG key between this slave and the master
key "internal-2-ipamworlwide.com." {
        algorithm hmac-sha256;
        secret "itPaWJ6xSOqNn2no/SZ2ex5nXZIJF3dUj1OXtmG8FBA=";
};

# associate the TSIG key with the master to sign communications
server 172.16.30.5 {
    keys { "internal-2-ipamworldwide.com." ;};
};

# secure (or disable) the rndc control channel
controls {
    inet 172.16.50.6 port 953 allow { "mgmt-net"; }
        keys { "internal-2-rndc.ipamworldwide.com."; };
};

# configuration options to disable recursive query processing,
# provide a null response to version queries, accept notify's
# only from the master, ignore requests for zone transfers and
# to perform zone transfers from the master on its internal
# IP address
options {
    recursion no;
    version "";
    allow-notify {  key "internal-2-ipamworldwide.com."; };
    allow-transfer { none; };
    transfer-source 172.16.50.5;
};
```

NLnet NSD Example   An NLnet NSD server example configuration subset
follows. Note that NSD is an authoritative-only DNS server and cannot be configured

to perform recursion. The ACLs need to be defined on a per zone basis though a
common zone *pattern* can be defined and applied across several zones.

```
# sample NSD server configuration (version 4.1.11)
# server options below ignore version queries and constrain the
# control channel
server:
    hide-version: yes
remote-control:
    control-enable: yes
    control-interface: ip4
    control-port: 8952
    control-key-file: control_key_filename.file
# define a TSIG key
key:
    name: "internal-2-ipamworldwide.com."
    algorithm: hmac-sha256
    secret: "itPaWJ6xSOqNn2no/SZ2ex5nXZIJF3dUj1OXtmG8FBA="
# for each zone define its name and file and options including
# from where notify's are permitted (signed), to whom to issue
# zone transfer requests and on what outbound address to query
# for and receive zone transfers
zone:
    name: "ipamworldwide.com"
    zonefile: "db.ipamworldwide.com"
    allow-notify: 172.16.30.5 internal-2-ipamworldwide.com.
    request-xfr: 172.16.30.5 internal-2-ipamworldwide.com.
    outgoing-interface: 172.16.50.5
```

PowerDNS Authoritative Example Configuration   A comparable configuration for the PowerDNS Authoritative server follows. We define our TSIG key in our backend database via SQL.

```
$ insert into tsigkeys (name, algorithm, secret) values
('internal-2-ipamworlwide.com.', 'hmac-sha256',
'itPaWJ6xSOqNn2no/SZ2ex5nXZIJF3dUj1OXtmG8FBA=');

$ select id from domains where name='ipamworldwide.com';
7

$ insert into domainmetadata (domain_id, kind, content)
values (7, 'AXFR-MASTER-TSIG', 'internal-2-ipamworldwide.com.');
```

The pdns.conf file enables configuration of database connectivity parameters to enable PowerDNS to connect to the backend. Some DNS settings can also be defined within the pdns.conf file or on the command line when invoking the pdns binary.

```
allow-notify-from=172.16.30.5
allow-recursion=none
version-string=anonymous
```

The control channel for PowerDNS, pdns_control, can be secured by defining a secret using the –secret option on the pdns_control command.

Knot DNS Configuration Example   A sample Knot DNS configuration file with comparable configuration follows.

```
# sample Knot DNS server configuration (version 2.3.3)
# ignore version queries
server:
  version: none

# define a TSIG key
key:
  - id: internal-2-ipamworlwide.com.
    algorithm: hmac-sha256
    secret: "itPaWJ6xSOqNn2no/SZ2ex5nXZIJF3dUj1OXtmG8FBA="

#define a notify acl
acl:
  - id: accept_notify
    address: 172.16.30.5
    key: internal-2-ipamworlwide.com.
    action: notify

#sign all transactions with the master
remote:
  - id: master
    address: 172.16.30.5
    key: internal-2-ipamworlwide.com.

zone:
  - domain: ipamworldwide.com
    storage: /var/lib/knot/zones/
    file: ipamworldwide.com.zone
    master: master
    acl: accept_notify
```

ADDITIONAL DNS DEPLOYMENT VARIANTS

Beyond the four basic trust sectors, additional sectors may be deployed to provide finer granularity partitioning of your DNS infrastructure.

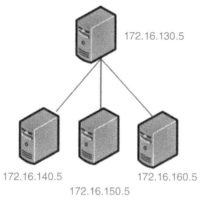

eng.ipamworldwide.com zone
20.172.in-addr.arpa zone
21.172.in-addr.arpa zone

Figure 5.6. Internal Namespace Delegation

Internal Delegation DNS Master/Slave Servers

In larger organizations, subdomains can be delegated to particular departments or divisions. Continuing our example, we've created a finance.ipamworldwide.com domain as nondelegated. This means that the configuration and resource records associated with the finance.ipamworldwide.com domain is included in its parent zone file for ipamworldwide.com.

For other departments, separate DNS administrators may desire to manage their own domain information. Let's consider an example. If the engineering department desires to run DNS for the eng.ipamworldwide.com domain, the team managing the internal second-level domain, ipamworldwide.com, the parent of eng.ipamworldwide.com may allocate a new delegated domain, that is, zone. The NS and glue records for the DNS servers authoritative for this new zone need to be configured on both the authoritative servers themselves and on those authoritative for the parent zone, ipamworldwide.com. Technically, the eng.ipamworldwide.com zone is authoritative for these NS (and glue) records, not the parent, but the parent must configure them to provide referrals down the domain tree.

Figure 5.6 depicts the example server deployment. As you can see, it looks exactly like that of the internal SLD DNS servers with a master and several slave servers. This is the common deployment configuration of authoritative DNS servers.

Multi-Tiered Authoritative Configurations

In some cases, it is desirable to add a third layer to supplement the two-tier master–slave model. This upper tier features a master DNS server, perhaps master

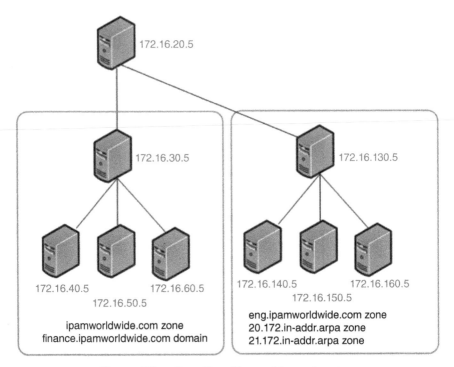

Figure 5.7. Three-Tiered Internal Server Structure

for all internal namespace which can in effect provide the true master database of DNS information for an organization. This scenario is illustrated in Figure 5.7.

This deployment scenario makes most sense when the zones served by these servers fall within the same administrative authority. If the eng.ipamworldwide.com zone had been delegated as a child zone of the ipamworldwide.com zone, it's best to keep master servers independent to minimize the possibility of cross-domain configuration issues.

Let's call this top-level DNS server a Tier 1 server. It is configured with all zones as type master. Our former master servers, 172.16.30.5 and 172.16.130.5, which we'll refer to as Tier 2 servers, are now configured as slaves, pulling zone transfers from our Tier 1 master. The set of original slave servers, at Tier 3, remain as such and continue to pull zone transfers from their respective Tier 2 servers. These Tier 2 servers, though slaves, are configured within the `masters` statement of each Tier 3 server's zone statements. Thus, no changes are required in the configuration of our Tier 3 servers. However, our Tier 2 servers must be modified as slave for each configured zone with the Tier 1 server identified as each zone's master. The Tier 1 server is referred to as the *primary master* in this configuration, as this is the server on which zone updates may be made directly, with zone transfers to Tier 2 and Tier 3 successively to update all authoritative servers accordingly.

Hybrid Authoritative/Caching DNS Servers

It may be tempting for some organizations to publish internal authoritative zones on recursive servers to perform dual-purpose resolution. In this scenario, an "authoritative recursive" DNS server would attempt to answer the query authoritatively, and failing that, perform recursion. The server would then cache resolution information received during the query resolution process, on behalf of resolvers.

For small to modest-sized organizations with a handful of internal DNS servers, this configuration works fine but does increase vulnerabilities as these multirole functions will be vulnerable to attacks for all roles. This trade-off between the economy of deploying fewer servers against the increased vulnerabilities of multifunction server deployments is one that must be assessed as you quantify risks and likelihoods to determine if the risk is acceptable.

Stealth Slave DNS Servers

Stealth slave DNS servers are so called due to the lack of NS and glue records for the server in the parent zone as we discussed in the case of hidden masters. Hence, you can hide masters or slaves by not publishing corresponding NS records. When traversing the domain tree, other DNS servers will not query this hidden server for resolution, as it will not be "advertised" in the parent's zone's referrals.

This type of configuration may be deployed for a slave server in order to reduce inter-server traffic or to control such traffic to a fixed combination of resolvers and other servers. Other than removing the stealth slave server's NS and glue records, the configuration is equivalent to that of a normal slave server.

Internal Root Servers

In environments where Internet access for internal clients is limited, prohibited, or otherwise generally unavailable, a set of internal root servers can be deployed to authoritatively resolve or more likely, refer queries otherwise not resolved by local recursive servers. In such an environment, we can eliminate the Internet caching servers since Internet queries are not permitted.

These internal root servers effectively replace the Internet root servers in the organization's DNS servers' hints files. In other words, these internal root servers are the ultimate authority and last chance for name resolution for internal clients. Thus, this configuration eliminates reliance on Internet name servers, but limits resolution to that information contained within the root servers and their delegated domain servers.

In this configuration, we can use the same configuration we specified previously in the Internal Authoritative DNS Servers section. But we need to define the hints file (root-hints.txt in our case) with a listing of the internal root servers instead of the standard Internet root servers. The hints file contains only NS and glue records for the root servers. If we use a simple example of three root servers, our referenced hints file might look like*

```
.                              IN NS
root1.ipamworldwide.com.
root1.ipamworldwide.com.   IN A 172.16.1.1
                           IN AAAA 2001:db8:4af0:f1::1
.                          IN NS
root2.ipamworldwide.com.
root2.ipamworldwide.com.   IN A 172.18.1.34
                           IN AAAA 2001:db8:4af0:a::1
.                          IN NS
root3.ipamworldwide.com.
root3.ipamworldwide.com    IN A    10.251.0.5
                           IN AAAA 2001:db8:4af0:c001::1
```

The configuration file on each of the root servers might look like the following:

```
acl internal-nets { 10.0.0.0/8; 172.16.0.0/12; 2001:db8:4af0::/48; } ;
options {
  recursion no;                    // iterative queries only
  allow-query { internal-nets; }; // allow from internal nets
  allow-notify { none; };          // disallow notify processing
  allow-transfer { none; };        // disable zone transfers
  allow-update { none; };          // disable updates
};

zone "." {
  type delegation-only;
  file "db.dot";
};
```

Each root server is a delegation-only type server, as denoted within the root zone declaration block at the bottom of the example configuration file above. The delegation-only type is a special form of type master that only responds with referrals, not answers. A multi-master deployment of internal root servers is possible for static zones like this that change infrequently.

Any modification to the root zone implies a new or modified top-level domain assignment and must be made by updating the db.dot file on each root server. There are no dynamic updates, notify's, or zone transfers. All changes must be made by

---

* Technically we could define our "root" as a private ".com" not "." if our internal space resides beneath the com TLD in its entirety though perhaps .local is more appropriate. This would obviate the need for another layer of servers and queries.

administrator modification of the db.dot file and requires a coordinated loading of the modified zone file on all masters to place it into service synchronously.

The following illustrates a portion of the example db.dot file, which contains resolution (delegation) data for the internal root zone.

```
$TTL 1d
. IN  SOA dns1.ipamworldwide.com. dnsadmin.ipamworldwide.com (
                  1 2h 30m 1w 1d );

//Refer queries to ipamworldwide.come and below to these servers
ipamworldwide.com.     IN NS  dns1.ipamworldwide.com.
                       IN NS  dns2.ipamworldwide.com.
                       IN NS  dns3.ipamworldwide.com.
                       IN NS  dns4.ipamworldwide.com.

//Refer queries to partner.net and below to these servers
partner.net            IN NS  dns-par1.ipamworldwide.com.
                       IN NS  dns-par2.ipamworldwide.com.
. . .

dns1.ipamworldwide.com.     IN A  172.16.40.5
dns2.ipamworldwide.com.     IN A  172.16.50.5
. . .
dns-par1.ipamworldwide.com.     IN A     172.16.199.2
dns-par2.ipamworldwide.com.     IN A     172.16.199.3
```

Referrals to other DNS servers we've configured previously enable authoritative resolution of queries. Here, any queries falling within ipamworldwide.com, including eng.ipamworldwide.com, will be sent to our internal authoritative servers. We would not include the 172.16.20.5 and 172.16.30.5 servers in this list as these are hidden servers.

For queries requiring external resolution, configure partner extranet DNS servers in the hints file as well to refer to internal partner-facing servers. In our example above, access to the partner.net domain and subdomains would be referred to authoritative DNS servers dns-par1.ipamworldwide.com or dns-par2.ipamworldwide.com. These servers may be configured as stub servers for the partner.net zone; alternatively, direct referral to the partner.net DNS servers may be used on these entries within the root zone file. The bottom line is that these root servers can delegate top-level domains to other DNS servers, which must in turn be configured to refer or resolve authoritatively for the corresponding domains and subdomains.

Deploying DNS Servers with Anycast Addresses

Configuring DNS servers with anycast addresses enables multiple DNS servers to utilize a common IP address. An anycast address is an address assigned to

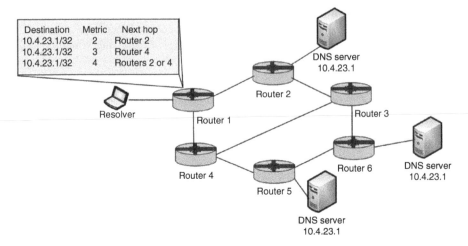

Figure 5.8.   Anycast Routing Table Example

multiple interfaces, typically on different nodes. Anycast is used when attempting to reach any one of the anycast addressable hosts without caring which host is reached. The routing infrastructure handles routing metric updates to track reachability and routing to the nearest host configured with the destination anycast address. Figure 5.8 illustrates an example with three DNS servers configured with anycast address 10.4.23.1.

While Router 1 has several routes to anycast address 10.4.23.1/32, Figure 5.8 depicts the three shortest routes to anycast address 10.4.23.1/32, corresponding to each of our three servers. The closest server is that homed on Router 2 and is two hops from Router 1. The next closest server is homed on Router 5 and is reachable in three hops via Router 4. Lastly, the server connected to Router 6 is reachable in four hops via either Router 2 or 4. The logical view from Router 1's perspective is illustrated in Figure 5.9, where the anycast IP address is considered a single destination, reachable via multiple paths.

Anycast Benefits   Deploying anycast provides a number of benefits.

- Simplified resolver configuration
- DNS server deployment flexibility by geography
- Improved resolution performance
- High availability DNS services
- Resilience from DNS denial of service attacks

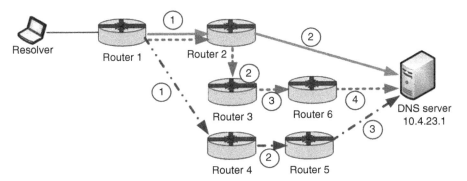

Figure 5.9.   Logical Routing Perspective from Router 1 Showing Hop Counts

Resolvers configured with the DNS servers' anycast address would have their queries routed to the nearest DNS server configured with that anycast address. Thus, regardless of where the resolver host connects to the network, the same anycast IP address may be used by the resolver to locate a DNS server. This localized query process can also improve performance of the resolution process. A query to a DNS anycast address should be routed to the closest DNS server, thereby reducing the round trip delay portion of the overall query process.

The outage of a DNS server can be communicated (by absence of communication) to the routing infrastructure in order to update routing tables accordingly. This requires the DNS server run a routing daemon using the routing protocol of choice to communicate reachability to the local router. Participation in routing protocol updates enables the local router to update its routing table with an appropriate metric and to pass this on to other routers via the routing protocol. Depending on the deployment of the DNS server, internal or external, a corresponding interior or exterior routing protocol would need to be running on the DNS server. The server simply needs to communicate that its anycast address is reachable. This is typically performed by assigning one of the server loopback addresses* as the anycast address and running a routing daemon on one or more ports advertising reachability to the anycast address. It would be especially useful if this routing update was linked to the status of the DNS daemon or service on the server, though application status is not generally considered when communicating IP address reachability.

Deploying anycast affords mitigation against denial of service attacks as evidenced by the DDoS attack on multiple root servers on February 6, 2007 (40). Of the six root servers targeted, the two most severely affected were those which had not yet implemented anycast. The other four root servers, having deployed anycast, enabled

---

* The term "loopback address" here refers to the software loopback address commonly implemented in routers and servers as the "box address" reachable on any of its interfaces.

the spreading of the attack across more physical servers. Thus a DDoS attack on the I-root server, which did not have anycast in place, severely impacted the ability of the server to respond to legitimate queries, while the attack on the F-root server, which had over 40 servers sharing the F-root anycast address, distributed the impact of the attack across these servers. This form of load sharing enabled the F-root server(s) to continue processing legitimate queries while suffering a barrage of artificial requests. That particular attack reached a peak of 1 Gbps, but with attacks today topping 1 Tbps, a broad anycast spread is recommended to reduce the impact of ever escalating attack bandwidth.

Anycast Caveats   While anycast provides many benefits, consider the constraints and caveats of deploying anycast. Because resolvers may query any DNS server configured with the anycast address at a given time, it's important that the resolution information configured on the server be consistent. For example, the implementation on Internet root servers consists of a set of master servers with static information. These root servers do not accept dynamic updates. If anycast is desired for dynamic zones, then each server must have a unicast address in addition to its anycast address*. This enables updates to be directed to the master's unicast address, which may in turn notify its slaves via their respective unicast addresses. A hidden master configuration can be used with the slaves configured with anycast addresses.

Another consideration is the requirement to run a route daemon on your DNS servers configured with anycast addresses. While routing of packets to anycast addresses is primarily a routing function, the unreachability of a DNS server host may result in lost query attempts. Such would be the case if static routes are used to configure routers with fixed metrics for the DNS servers configured with a common anycast address. Should a server become unavailable, the serving router has no way to detect this and would not re-route packets destined for the anycast address. Therefore, incorporating a routing daemon on the DNS server improves overall robustness. Should a server fail, the local router will determine that it is no longer reachable and will update its routing table and those of other routers via routing protocol updates. Internet root servers support BGP, given their deployment on the global Internet, though deployment within organizations will likely require support for OSPF, IGRP, or the interior routing protocol of choice.

Lastly, troubleshooting is a little more challenging when using anycast. Debugging a bogus response from a server's anycast address is difficult given the server ambiguity. To identify which anycast-addressed server is troublesome, it's a good idea to configure the server identification with the NSD's or Knot's `identity` parameter or BIND's or PowerDNS' `server-id` parameters. You can define a string identifier or just use the hostname argument to use the server's hostname. This value is retrievable by issuing a query with qname="ID.SERVER", qtype=TXT

---

* Every anycast server will require a unicast address for administration, but to support dynamic zones, an additional unicast address is required to provide an interface for updates, notify's, and zone transfers.

with qclass=CHAOS. Using the dig utility, this looks like dig id.server chaos txt @<anycast-address>.

Perhaps a better way is to use the dig +nsid argument with a query so you can correlate a bad response with the server identification in one transaction.

```
dig +nsid <query> @<anycast-address>
```

Configuring Anycast Addressing   Participation in routing protocol updates is key to maximizing the benefits of your DNS anycast implementation. The first step is to assign your chosen anycast address to your set of DNS servers sharing the address. This address is commonly assigned to the server loopback address. On Linux or Unix systems, this can be performed using the ipconfig command such as follows:

```
ifconfig lo:0 10.4.23.1 netmask 255.255.255.255
```

The simplest form of routing participation is to apply static routes which merely tells the local router to route the anycast address to the interface connected to the DNS server, and then to redistribute this static route among its peer routers.

```
ip route 10.4.23.1 255.255.255.255 172.20.23.1
```

But the use of static routes offers no benefits of rerouting based on routing protocol metrics such as network load, not to mention server reachability as just discussed. Use of a route daemon on the DNS server itself will enable the server to provide reachability information dynamically such that performance from a routing perspective among all anycast servers is optimized.

The open source routing package, quagga is commonly used for this purpose. Quagga supports a variety of routing protocols including RIP, RIPng, OSPFv2 and v3, and BGP. Configuration of quagga is relatively simple and entails configuring the zebra kernel routing manager component of quagga to configure it to include the server's network interfaces such as

```
interface eth0
ip address 172.20.23.1/24
interface lo
line vty
```

The virtual teletype (vty) interface enables interactive access to the configuration while zebra is running. With zebra runnning as the interface to the kernel, configure the corresponding routing protocol configuration. For example, to run OSPF, the ospfd.conf file should include interface and OSFP declarations.

```
interface eth0
interface lo
router ospf
   ospf router-id 172.20.23.1
   log-adjacency-changes detail
   redistribute connnected
   network 172.20.23.0/24 area 0.0.0.100
line vty
```

After configuring quagga's zebra component and the routing component for your routing protocol, you then need to update the connected router configuration to associate the router interface with the assigned OSPF area, 100 in this case.

OTHER DEPLOYMENT CONSIDERATIONS

High Availability

We've implicitly addressed this already throughout this chapter but it bears repeating. Deploy multiple servers within each trust sector and size the servers such that the query load can be handled by a subset of the total number of servers should one or more become unavailable.

Multiple Vendors

Deploying multiple DNS servers running software from diverse vendors can help protect against the widespread impact of an attack exploiting a given vendor's implementation. For example, within your recursive server trust sector, you may choose to deploy one set of DNS servers running ISC BIND and another set running Unbound. This complicates your server management processes but this diversification approach can help deflect vendor-specific targeted attacks.

Sizing and Scalability

Within each trust sector, you must also size the quantities, locations, and server specifications required to achieve high availability and resiliency. One logical location deployment approach is to place servers in more densely populated regions of your network to provide performance by proximity. However, sufficiently sized network links with redundancy enables a more centralized approach if preferred.

Server specifications need to be based on expected query load and resiliency requires the addition of one to a few servers to a design that fully meets the query load; the intent is to ease the load per server while enabling the outage of a server or two (or links thereto) to not hinder overall query performance. Monitor and measure your servers' performance from a query rate, CPU, memory, and I/O perspective to track and assure adequate performance; monitoring is also critical for detecting threat events as well.

Load Balancers

Load balancers may be deployed in front of a set of DNS servers to enable the processing of DNS queries across a number of DNS servers. DNS queries would be addressed to the load balancer IP address and the load balancer would be responsible for forwarding the query to an available DNS server for resolution. Load balancers introduce an additional element within your network and DNS flow, but can be instrumental in providing improved performance and resiliency.

Note that the load balancer should pass through the query source IP address from the original query instead of NATing this address. This will prevent bypassing of your address-based ACL configurations on your DNS servers, unless you configure comparable ACLs on your load balancers.

Lab Deployment

A laboratory network should be deployed which is physically and networking-wise separate from your production network. While full replication of your production network within a lab is likely unrealistic, a subset of your deployment in terms of modeling each DNS server vendor and version as well as role can help you test network changes in a controlled environment without affecting the production network should a test configuration fail.

PUTTING IT ALL TOGETHER

We've presented DNS deployment scenarios in this chapter in the form of trust sectors while addressing a wide variety of configurations, each targeted at addressing a specific DNS resolution purpose or role. This approach enables segmentation of DNS resolution traffic and application of corresponding DNS server security controls. Depending on the size and scale of your IP network, you may choose to implement several building block scenarios for different applications on your network. Just remember that there are no true cookie-cutter answers; each of these scenarios should be evaluated based on your individual needs.

Most organizations will deploy at minimum the External DNS, Internal DNS, and Recursive DNS trust sectors. Those with partners may also add the Extranet DNS sector covering both inbound and outbound query resolutions. Other trust sector variants may be implemented as well. Figure 5.10 illustrates one possible incarnation of a holistic DNS infrastructure.

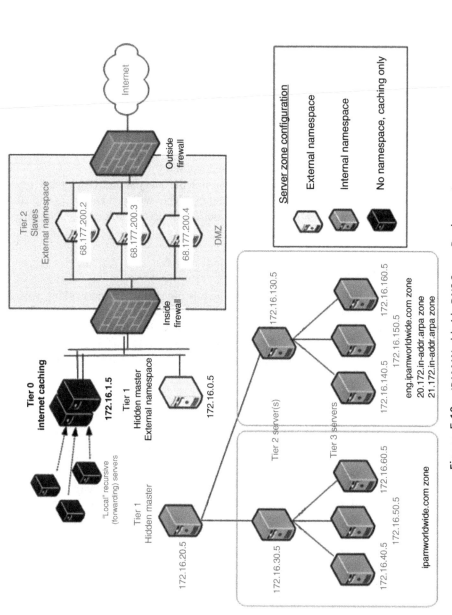

Figure 5.10. IPAM Worldwide DNS Server Deployments

6

SECURITY FOUNDATION

INTRODUCTION

This chapter focuses primarily on securing the host hardware on which your DNS components run. DNS hardware and software components, deployed with a trust sector architecture in mind, provide the foundation for your DNS services. In this chapter, we'll introduce core security activities, including risk assessment, data security, threat detection, and event recovery. These functions apply to this and the ensuing five chapters to address specific DNS threats and vulnerabilities. Hence, we will mention references to future chapters where appropriate though our main focus in this chapter lies with DNS server hardware and software, including component operating systems, kernel, and DNS application components.

Most organizations define and enforce common hardware and operating system security policies for end user devices and analogous policies for IT systems such as DNS servers. Application of such policies to DNS servers enables uniform monitoring and enforcement at the hardware and operating system level. DNS application level policies are also needed to monitor, detect, and remediate DNS level attacks.

DNS Security Management, First Edition. Michael Dooley and Timothy Rooney.
© 2017 by The Institute of Electrical and Electronic Engineers, Inc. Published 2017 by John Wiley & Sons, Inc.

HARDWARE/ASSET RELATED FRAMEWORK ITEMS

Protecting your DNS servers is foundational to fortifying multiple layers of security within your defense in depth strategy. This includes physical security, host security, and data security. As such, we'll continue to apply the NIST cybersecurity framework in this chapter as a baseline, which we will frame in the context of DNS. The following sections summarize the relevant NIST cybersecurity framework categories that we will discuss in the context of selected hardware, operating systems/kernel, and DNS application software.

An example full set of framework desired outcomes related to DNS security can be found in Appendix A. This appendix also includes a sample framework core profile with references to related security standards which you may access for additional details. Note that your particular implementation may require addition, modification, or deletion of some of the tasks discussed.

Identify: Asset Management

Asset management at base requires knowledge and documentation of the assets under management. In this case, it also entails asset prioritization in terms of level of vulnerability versus network or business criticality.

Physical Inventory (ID.AM-1) The first ID.AM subcategory stipulates that physical devices and systems within the organization are inventoried. The following tasks should be endeavored to attain outcomes specified by this subcategory. A centralized, controlled document or repository should be maintained to document the inventory of DNS components by role, site, asset information, contact name, and related asset tracking information.

- Document inventory of recursive DNS servers
- Document inventory of internal authoritative DNS servers
- Document inventory of external authoritative DNS servers
- Document use of external DNS providers for recursion or authoritative DNS
- Document inventory of resolvers (clients, stub resolvers)

Software Inventory (ID.AM-2) This subcategory stipulates that software platforms and applications within the organization are inventoried. This should include DNS software used for each DNS server you manage, which may be the same vendor for all or you could employ different vendor software for different roles; for example, Microsoft DNS for internal authoritative DNS and Linux NSD for external DNS. Track the current version and patch level for each component as well. Resolver software running on end user and other unmanned devices that use DNS must also be inventoried. The following documentation must be maintained to achieve outcomes associated with this subcategory.

- Document DNS vendor and software version for recursive DNS servers
- Document DNS vendor and software version for authoritative DNS servers
- Document DNS vendor and software version for external authoritative DNS servers
- Document resolver vendor and version of resolvers (clients, stub resolvers)

External System Inventory (ID.AM-4) If you're using a DNS service provider for recursion or hosting services, document such use and affiliated information. Use of external recursive servers generally entails pointing resolvers to an anycast service provider IP address, which should be documented, links to the service provider policies and vulnerability notifications should be documented and vulnerabilities monitored. If you use a service provider to host your external namespace, document service provider DNS information related to service level agreements, security monitoring and notification policies, and domains managed.

- Documentation of external DNS provider policies is consistent with enterprise security policy
- External DNS system services for recursive and authoritative services comply with organizational and relevant regulatory security requirements
- External information systems are catalogued

Resource Prioritization (ID.AM-5) This subcategory states that resources (e.g., hardware, devices, data, and software) are prioritized based on their classification, criticality, and business value. Assets of relatively higher value to an organization would likely warrant relatively higher priority in terms of application of security controls. This task requires review of your itemized hardware and software inventory to rank in priority order.

- Categorize and prioritize DNS servers accordingly; plan for contingencies for potential compromise that could affect your ability to resolve DNS (recursive, resolver compromise) and that which could redirect external end users from your web servers (authoritative compromise).
- Categorize DNS software vendor supply chain risk. Consider whether you are using one vendor's DNS solution or have diversified.
- Define contingencies for DNS server failure or failure of external DNS provider(s).

Identify: Business Environment

Critical Services Resilience (ID.BE-5) This subcategory recommends that resilience requirements to support delivery of critical services are established. DNS is certainly a critical service, so resiliency of the service must be designed in. While this subcategory could apply more broadly to your deployment approach in terms of deploying multiple site-diverse servers, hardware resiliency options are widely available which can serve to provide intra-node redundancy. Such options include multi-processor systems with dual power supplies, RAID hard drives, and even hardware clustering. Both intra- and intersite redundancy implementations should be considered for more critical components such as external master DNS servers, for example.

- Based on DNS server prioritization, deploy hardware redundancy potentially including multiple servers in multiple sites, hardware clustering, anycast, load balancers, and/or IP virtual server (IPVS)
- External DNS services should be contracted to include high availability SLAs for external DNS and/or to supplement an in-house external DNS server deployment
- If you rely solely on external DNS service providers, consider contracting with a second service provider for added resiliency should one service provider come under attack

Identify: Risk Assessment

This set of subcategories specifies the process for identifying and maintaining a threat list, and for applying likelihood, business impact, and risk assessment.

Document Vulnerabilities (ID.RA-1)   The first subcategory within the risk assessment category stipulates that asset vulnerabilities are identified and documented. This requires researching vulnerabilities for the systems you have deployed and continually monitoring vulnerability reporting sites as well as your vendors' sites for vulnerability and patch announcements and applying them accordingly in a timely manner. This applies up the stack from each DNS component.

- Document and track vulnerabilities for DNS server hardware components
- Document and track vulnerabilities for DNS server kernels
- Document and track vulnerabilities for DNS server operating systems
- Document and track vulnerabilities for DNS server software applications
- Document and track vulnerabilities for DNS resolver software applications on end user and other devices
- Document and track vulnerabilities for the DNS protocol, including those we'll discuss in Chapters 7–11
- Document risks for human errors in server misconfiguration, missed detections, false positives, and so on
- Document risks based on DNS server location deployment with respect to site security and environmental risks such as earthquakes, tornadoes, volcanoes

Seek Multiple Vulnerability Report Sources (ID.RA-2)   Threat and vulnerability information should be sought from multiple reputable sources, including information sharing forums and sources. We recommend you subscribe to the US Computer Emergency Readiness Team (US-CERT) email list (41).

- Actively monitor threat and vulnerability information sources like US-CERT (42), CVE (43) (which is sponsored by US-CERT), and related reputable web resources
- Collaborate with other security personnel especially within your industry to promote open sharing of threats, vulnerabilities, and mitigations
- Subscribe to security feeds offered by each vendor who produces the operating systems, operating system utilities or daemons, and applications (DNS) software in use within your network.

Document Threats (ID.RA-3)   Based on your threat research for components in use for your DNS services, document each threat, both internal and external. This master list will be a living document as new threats are identified over time and will serve as the enumeration of threats to your network and DNS in particular.

- Document and track threats for DNS server hardware components
- Document and track threats for DNS server kernels
- Document and track threats for DNS server operating systems
- Document and track threats for DNS server software applications
- Document and track threats for DNS resolver software applications on end user and other devices
- Document and track threats for the DNS protocol, including those we'll discuss in Chapters 7–11
- Document threats related to personnel issues as well as natural and unnatural disasters

Identify Threat Impacts (ID.RA-4)   Analyze each threat you've documented to assess the potential business impact and likelihood of each. Just as your list of documented threats is a living document, so too is the associated threat impact. For example, certain weather related threats may be seasonal; some newly identified attack threats may be more prevalent for certain types of organizations or websites or impose a higher business impact severity, etc. And as you implement controls against certain threats, a residual threat will likely remain, albeit hopefully of lower impact.

- For each identified threat and vulnerability, document the likelihood of the occurrence.
- For each identified threat and vulnerability, document the related business impacts should the threat materialize.

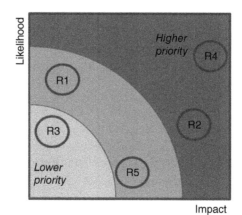

Figure 6.1.   Risk Likelihood-Impact Plot

Determine Risk (ID.RA-5)   Given the prior subcategories, ID.RA-1-4, you should possess a document enumerating threats, vulnerabilities, likelihoods, and impacts for your network and DNS. Based on your analysis, you can determine the relative risk of each. You could graphically depict the risk of each threat similar to our example risk likelihood impact plot we introduced as Figure 1.3 in Chapter 1, reproduced for convenience as Figure 6.1.

- Assess the risk of each identified threat and vulnerability based on the likelihood and related business impacts of the occurrence.
- Iterate your analysis to consider the relative risk for higher priority assets, for example, DNS servers, as identified in subcategory ID.AM-5. A given threat may have a higher impact when imposed on a master DNS server versus a slave, for example.

Protect: Access Control

Moving into the protect function, access control to DNS components is our first category.

Identities and Credentials (PR.AC-1)   For all DNS components, authorized user identities and credentials need to be managed. The identities of users authorized to access each component should be documented. Authorized users with a job function and scope of responsibility necessitating server access should be permitted access. Access permissions (authorization) should be scoped to the extent possible only to those functions necessitated by the user role and identity. User activity should be audited periodically and user identities should be removed for those users, whose job function no longer requires access, including those departing the group or organization. Credential policies such as password length, character composition, change frequency, lockouts for your organization should apply to those authorized to access DNS components.

Physical Access Controls (PR.AC-2)   Access control subcategory two states that physical access to assets is managed and protected. Deployment of physical and/or virtual DNS servers within secured datacenters is a good first step. Only staff with a job function and scope of responsibility necessitating datacenter access should be permitted access; physical access should be controlled via badge reader or similar electronically trackable mechanism for auditing. A process for removal of servers from the datacenter with requirements for documentation and approval enables proper authorization and communication to datacenter staff regarding conscious removal of assets. This of course feeds back to the physical inventory subcategory as well, ID.AM-1.

Remote Access Controls (PR.AC-3)   Policies should be enforced to secure remote access to DNS servers for administration purposes. Only staff (internal or contractors) with a job function and scope of responsibility necessitating remote access should be permitted access. Login/password authentication should be required and communications should require an encrypted connection. Command sets should be restricted if possible to only those required by each user.

Access Permissions (PR.AC-4)   Access permissions are managed, incorporating the principles of least privilege and separation of duties. For each user identity, the level of access should provide only those commands or functions required of that user to perform his or her job function.

Protect: Data Security

Security of data stored on or communicated to or from a DNS component in this context applies to device kernel, operating system, and DNS application data. We'll discuss security of DNS configuration data, including resource records in ensuing chapters with respect to defending against poisoning of DNS cache or authoritative configuration data.

Data at Rest (PR.DS-1)   Data at rest refers to data residing on a server, such as kernel and operating system configuration, including configured services, ports, users, partitions, the file system, and file contents. Change control procedures should be documented and followed with respect to timing and content of planned changes. Periodic backups and storage of this data offsite provides a fallback for restoration in the event of server failure. Data stored offsite must be transported and stored securely via encryption. These processes should be applied to DNS application data, configuration information, zone repositories, and cached or journaled (pending updates) records as will be discussed in ensuing chapters.

Data in Transit (PR.DS-2)   The second aspect of data security relates to data in transit. As we've discussed in the opening chapter of this book, DNS necessarily supports several transit routes among various DNS and non-DNS components for name resolution as well as for configuration and updates. Most of the vulnerabilities discussed in the following five chapters relate to DNS data in motion and will be discussed accordingly.

Securing data in transit implies encryption or at least digital signing, and these techniques can be applied to most transit paths, though such application requires implementation and maintenance effort, for example, DNSSEC. But ongoing effort is a worthy price for security! Other paths, such as from stub resolvers to recursive servers or forwarders are less often secured but the recent introduction of DNS cookie support in ISC BIND (9.10.3), for example, provides a step up from zero connection security. DNS cookies provide a lightweight DNS message authentication mechanism (24). Table 6.1 summarizes the basic DNS data flows from Figure 1.2 and associated communications' security mechanisms.

Asset Movement (PR.DS-3)   This subcategory stipulates that assets are formally managed throughout removal, transfers, and disposition. All DNS servers and resolvers, physical or virtual, must be inventoried per ID.AM-1, and any movement into, out of, and within the inventory listing must be controlled per this subcategory.

TABLE 6.1   Basic Data in Motion Controls

Transit Path	Transit Endpoints	Key Security Mechanisms
Recursive query	Stub resolver–recursive DNS server	ACLs, DNS cookies, TCP, DNSCrypt; DNSSEC possible but rarely implemented on stub resolvers
Iterative query	Recursive DNS server-referral and authoritative DNS servers	DNSSEC
Dynamic update	IPAM systems, clients, DHCP servers, and authoritative DNS servers	ACLs, transaction signatures (TSIG, GSS-TSIG)
Zone transfers	Master DNS server and slave DNS servers	ACLs, transaction signatures (TSIG)
DNS configuration	IPAM system, file editor or transfer to and DNS server	Secure shell, secure copy, secure FTP, transport layer security

This means that any proposed addition, movement, or removal of DNS components must be documented, reviewed, and approved by involved parties prior to commencement.

Data Leaks (PR.DS-5)   The requirement to protect against data leaks may seem irrelevant to a service whose very purpose is to freely divulge answers to queries and to indicate if no such answer exists. But that's not the "data" of concern when addressing leaks. Controls must be implemented to prevent leakage of information regarding the DNS deployment, IP addresses, server implementations and versions, server credentials, security policies, and related details that may provide information to would-be attackers.

Such information can be leaked through DNS reconnaissance, server hacking, traffic snooping, phishing, or through social engineering methods such as posing seemingly innocuous questions to unsuspecting staff members related to this information. Protection requires technical and staff training controls to reduce the risk of data leaks.

Integrity Checking (PR.DS-6)   Integrity checking mechanisms are used to verify software, firmware, and information integrity. Checksums on configuration files and digital and transaction signatures can provide a validity check regarding software or configuration file and DNS transaction integrity.

Protect: Information Protection

Baseline Configuration (PR.IP-1)   A baseline configuration of network systems, including DNS servers and host systems with stub resolvers should be created and maintained. Such "standard build templates" provide consistency through uniformity which helps constrain the breadth of software installed on devices of various types; therefore, vulnerabilities introduced through the use of nonstandard software or hardware are reduced. For end user systems, the resolver software should be defined as included with the build, which almost always consists of that supplied with the corresponding device operating system. For DNS servers, the list of hardware and software components, including operating systems, DNS software, security, monitoring, and auditing utilities and so on, should be documented.

Change Control (PR.IP-3)   The baseline configuration should be a controlled document, meaning that any additions, changes, or deletions are proposed, reviewed, approved, and communicated among relevant parties. Changes to the configuration of the authorized software, for example, the DNS configuration, should also be planned, reviewed, approved, and staged.

Operating Environment (PR.IP-5)   Your organization should document and maintain policies and regulations regarding the physical operating environment for organizational assets such as DNS servers. This includes providing emergency power shut-off, fire protection, temperature and humidity controls, water damage protection and server room or datacenter access controls and auditing.

Protect: Maintenance

Maintenance and Repair (PR.MA-1)   Maintenance and repair of DNS servers should be performed and logged in a timely manner, with approved and controlled tools. Maintenance and repairs should use preapproved documented tools and processes. Any repairs requiring support from outside personnel, for example, vendor staff, should be pre-authorized and outside personnel should sign in and out and be escorted at all times by an authorized organization team member. Repairs requiring removal of a DNS component must be approved along with associated contingencies and the component sanitized of any sensitive information such as user accounts.

Remote Maintenance (PR.MA-2)   Any DNS server maintenance or repair performed remotely must be preapproved and a remote connection opened for the duration of the activity. Strong credentials for remote access are needed for identity verification. Remote access is to be logged as are DNS server commands and diagnostic actions.

Removable Media (PR.PT-2)   Any removable media must be protected and its use restricted according to policy. Vendors may supply software updates via USB, DVD, or other removable media format. Configuration and backup information may also be copied to removable media for backups. Swappable hard drives may contain configuration or sensitive information. Such media must be securely stored to prevent unauthorized access to the media and such media must be securely transported if necessary and must be sanitized prior to disposal by the removal of any sensitive information stored on the media.

Controlled Access (PR.PT-3)   This subcategory states that access to systems and assets is controlled, incorporating the principle of least functionality. Only those users whose job function and role necessitate access should be provided access. The breadth of functionality permitted for each user should be constrained to the extent possible by permission controls on the device. The functionality of the device itself should be constrained to the minimum functionality required to perform that device's role. Hence, for a DNS server, any non-DNS related services, except those necessary for diagnostics and auditing, should be removed. In addition, restrictions should be defined for unnecessary TCP/UDP ports, system or application files, processes, users, and the file system.

Detect: Anomalies and Events

The detect function covers the monitoring and detection of incidents which may comprise an exploitation of one or more vulnerabilities in the form of an attack, human error, or natural disaster.

Incident Response (DE.AE-5) An incident response plan should be established to define potential incidents and the corresponding response plan. The plan should be reviewed, approved, and updated periodically. Thresholds and alerts in monitoring systems should be established to detect and report potential security incidents, for example, for process and hardware states and well as I/O volumes. The response plan should also define incident analysis, containment, eradication, and recovery, along with the communication of status reporting.

Detect: Security Continuous Monitoring

DNS components must be continuously monitored to enable detection of potential incidents.

Environment Monitoring (DE.CM-2) Physical access control systems and the physical environment where DNS components are located should be monitored to detect potential incidents. Badge-in access should be required for access to critical infrastructure including DNS servers. Access logs should be reviewed. Surveillance systems should also be deployed and reviewed to detect "tailgating," the entry by an unauthorized person before the door closes, for example, and physical removal of assets.

Malicious Code Detection (DE.CM-4) Malicious code such as malware can infect any computing system. Preventive measures include DNS server hardening, network filtering, and detecting malware through scanning software or by detection of changes in system files. Configuring DNS response policy zones, aka DNS firewall, for your recursive servers enables detection of malware implanted on end user devices or other devices running stub resolvers. Please refer to Chapter 11.

Authorization Compliance (DE.CM-7) This subcategory discusses the monitoring for unauthorized personnel, connections, devices, and software. This requires monitoring for the incidence of unauthorized personnel in secure areas such as datacenters, connections to servers from unauthorized IP addresses, ports or credentials, any devices not specified within the asset inventory and any software installed on devices beyond that specified in the device baseline. Any such incident of noncompliance should trigger a notification for investigation in accordance with the incident response plan.

Vulnerability Scans (DE.CM-8)   Periodic vulnerability scans should be performed to detect new vulnerabilities and to verify deployed mitigation controls. Any new vulnerabilities or inadequate mitigation measures should be analyzed to assess overall relative risk based on likelihood and business impact and to define new or improved mitigation approaches for development and deployment.

Respond: Analysis

The response to detected incidents must be documented and agreed. The response plan document should be updated based on lessons learned as incidents are handled.

Incident Impact (RS.AN-2)   Upon incident detection, the incident should be analyzed to assess and understand the impact of the incident. The incident response plan should be followed and impacted groups involved in responding to contain, eradicate, and recover from the incident, while communicating status in accordance with the response plan. New information or lessons learned should be incorporated into an update of the response plan based on review, concurrence, and approval by appropriate members of the organization.

Forensics (RS.AN-3)   Forensic analysis on detected incidents should be performed to go beyond the symptoms of the incident to identify the ultimate cause and to enumerate those vulnerabilities exploited or attacked. This analysis is useful for identifying new or morphed attack vectors and vulnerabilities, and to qualify the effectiveness of any defensive controls that were intended to protect against such an attack.

Incident Categorization (RS.AN-4)   Incidents need to be categorized in a manner consistent with incident response plans. This is helpful in terms of prioritizing actions and inclusion of appropriate staff to analyze, contain, eradicate, and resolve the incident in a timely manner.

Respond: Mitigation

Incident Containment (RS.MI-1)   Incidents need to be contained in accordance with the incident response plan. Deployment of DNS components by function as we discussed in the prior chapter should provide containment to the corresponding trust sector. Further containment steps must be undertaken based on the incident itself to prevent broader impact on multiple DNS servers, resolvers, or other network systems.

Incident Mitigation (RS.MI-2)   As the incident is contained, contingency plans implemented, and forensics analyses conducted, mitigation approaches for the vulnerability that led to the successful incident should be defined, evaluated, agreed upon, and implemented. The vulnerability list, risk assessment, and incident response plan should be updated accordingly.

New Vulnerabilities (RS.MI-3)   Newly identified vulnerabilities need to be incorporated into the known vulnerability list. Such vulnerabilities may be detected through vulnerability scans, via vendor vulnerability announcements, from general security information sources, or from analysis of your DNS traffic. Each new vulnerability should be analyzed with respect to likelihood and business impact to define relative risk. Based on this assessed risk, the vulnerability should be proactively mitigated or documented as an accepted risk.

Recover: Recovery Planning

The recovery function seeks to restore systems and the network to normal operation through contingency plans and full restoration of affected elements.

Recovery Plan (RC.RP-1)   The incident recovery plan is executed during or after an event. During the event, contingencies and workarounds are put in place to restore service levels in the face of a disruption, compromise, or outage. After incident eradication, each affected system should be restored to prior operation and a known working state.

Recover: Improvements

A continuous improvement approach enables documented updates of response plans and strategies to more effectively respond to future incidents.

Update Response Plans (RS.IM-1)   After incident recovery, a postmortem discussion with involved staff is useful for reviewing the incident, possible defensive and mitigation steps to improve response, and recommended response plan updates to incorporate lessons learned.

Update Response Strategies (RS.IM-2)   Incident response strategies should be reviewed and updated as appropriate. For example, an update may necessitate involvement of certain groups earlier in the process.

Update Recovery Plans (RC.IM-1)   Recovery plans should also be updated to incorporate lessons learned.

Update Recovery Strategies (RC.IM-2)   Recovery strategies should be reviewed and updated should any improvements be borne out of the analysis of the incident recovery.

DNS SERVER HARDWARE CONTROLS

Most host- or device-based security controls you've implemented will likely serve as a solid base for those you should implement for your DNS servers. Applying the NIST Cybersecurity Profile, you can assess your current status with respect to this recommended security approach and define your desired future state to plan for improvements in your overall security. The controls we discussed in the prior section apply in particular to securing DNS servers. We'll summarize the key controls in this section with respect to securing your DNS hardware and software, and the following six chapters will discuss controls to defend against additional vulnerabilities of DNS components and of those that leverage DNS components.

Implementing host access controls provides a foundational layer of defense against attempts to attack a given hardware system, disrupt services running on the hardware, or to manipulate data or applications operating on the system. Each host operating system should be *hardened*, which entails securing the operating system such that it provides necessary file system components and permits only the level of permissions, users, daemons, and other resources required to perform the functions for which the hardware is intended as we'll discuss next. Additional security controls discussed later should be enforced to assure the availability and integrity of each physical or virtual server.

DNS Server Hardening

Hardening is the process of transforming a generic off-the-shelf server into a purpose-built DNS server. Defining single purpose hardware helps streamline server functions which reduces vulnerabilities in general – the less software running on the server, the fewer the vulnerabilities in general. Several vendors offer DNS appliances that are hardened to varying degrees. This offers you convenience though less direct control. The following controls should be considered for your DNS servers, assuming you're running on a Linux or Unix host.

Linux Kernel Security

- The kernel should not enable loading of drivers on the fly.
- The kernel should provide packet filtering and manipulation.
- The kernel should not include support for file systems not used directly for the purposes of DNS and supporting diagnostic, logging and auditing functions.
- The boot process should be controlled and should disallow interruption. This prevents attackers from interrupting the process and attacking the appliance during boot.
- The kernel should be built and configured in a secure environment, not connected directly to public networks.

DNS SERVER HARDWARE CONTROLS

- Various kernel settings need to be configured to protect against common networking attacks (i.e., route redirection, source route spoofing, ICMP redirects).

File System Security

- The file system should be pared down to include only necessary DNS and supporting binaries.
- System configuration files and logs have the strictest file permissions possible so as to not leak potentially useful information to an attacker.
- Run DNS services in a jailed environment. If your DNS daemon is compromised, the attacker may gain access to other files on the server. It is possible to create an alternative file system that mimics the default file system and locks the daemon inside of this mimicked file system structure making it more difficult for a compromised daemon to get outside of the default directory system and violate other system files. All major DNS implementations provide an installation or configuration option that creates this chrooted jail environment for you but may seem a bit odd because at least in Red Hat Enterprise Linux you end up with a file structure like this: /var/named/chroot/dev/etc/var/named. The root of the file system from the perspective of the jailed daemon is actually four layers deep into the file system and contains only a small, recreated subset of the real file system. For an example on how to do this manually see Boran's Running the BIND9 DNS Server Securely (44).
- The jailed environment must be sterile. Should an attacker infiltrate the jail, there is little else that can be done or attacked on the appliance or on other devices from the appliance.
- The DNS binaries and data files have no privileged attributes.
- DNS data files are controlled.
- Critical binaries are statically linked.

Process Security

- Run your DNS service as an unprivileged process. Services run as a particular user account on a given system and thus attain the access levels of that user. If your DNS daemon is running as root and the daemon is compromised, then the attacker has effectively taken control of a service with root privileges which gives him/her greater permissions on the system. In most Linux implementations running ISC BIND, the default installation procedure creates a user account called named and the named daemon runs as that unprivileged user account. Other DNS vendor software products are also configurable as such.
- DNS processes run in a sterile environment.
- DNS libraries are separate from system libraries.

Network Security

- The only open network ports are for use by DNS and diagnostic services.
- Other ports should operate in stealth mode, not responding at all to connection attempts.
- The console interface should be secured via an SSH connection.
- Firewalling can be configured via iptables. Any combination of networks, individual hosts, protocols – TCP and UDP, and individual ports can easily be placed into stealth mode so that the appliance will not respond to attempted connections.
- Denial of Service protection can be configured with iptables to rate-limit ingress TCP and UDP packets.

Additional DNS Server Controls

Depending on your operating system in use, the following areas should be addressed in your hardware controls. Other or alternative controls may be dictated by your particular network implementation and requirements.

- Monitor security advisories for the operating systems and DNS software that you use and assess the risk to your deployment. Apply security patches quickly based on respective severities and risk assessment.
- Remove or turn off system users and services/daemons and binaries not required for the server's intended function
- Implement host firewalls by turning off any IP addresses, protocols, and ports not required for the server's function or for management of the server
- Deploy applications required for each server's function in a "jailed" environment. See the preceding File System Security section for more details.
- Implement user access controls, changing vendor default passwords, requiring unique user identifiers and passwords per user; apply password schemas and aging policies.
- Review host/server access logs for anomalies or policy violations and conduct periodic penetration tests
- Train personnel regarding these policies and hold them accountable for adherence.

Physical Security Controls   Physical security controls for all network components including DNS protect against attackers gaining physical access to components for manipulation, disruption, or theft. Generally, your common physical controls should be adequate to protect your DNS infrastructure, provided you account for the following and potentially others depending on your organization's policies.

- Provide access controls potentially including security cameras for persons accessing your sites and particularly network and computing equipment rooms such as datacenters or phone closets.
- Require badging of personnel and contractors and provide sign in and temporary badging of visitors to the facility.
- Implement controls on publicly available wifi networks or network jacks to prevent access to unauthorized components.
- Classify computer media and paper documents containing sensitive information and control storage, accessibility, and disposal.
- Require documentation and approval for personnel access and the extent of permissible access.
- Review physical access logs for anomalies or policy violations and conduct periodic tests of access controls
- Train personnel regarding these policies including the questioning of unbadged or unrecognized persons on the premises.

Operations Controls Operations controls entail the repeatable standardization of processes and procedures, including assuring the awareness of relevant security policies and procedures across all personnel. This necessitates training of all personnel regarding security policies, how to identify potential violations and to whom such violations should be reported and tracked. Security procedures should be reviewed periodically, particularly after an incident detection–response–recovery cycle. The following summarize key operations controls:

- Document a security plan which is periodically reviewed and updated with lessons learned and technology updates. The overall security plan encompasses asset inventory details, implemented controls, detection mechanisms, incident response plans, and recovery plans. Define procedures as well as policies covering all aspects of vulnerability identification, asset protection, incident detection, response and recovery.
- Establish a training program to educate employees/contractors of policies and procedures and require acknowledgement from each team member.
- Garner management level approval and acknowledgement of policies and procedures by employees and contractors.

SUMMARY

This chapter has introduced key DNS security outcomes as related to the NIST Cybersecurity Framework Core and discussed security controls you should consider for protecting the hardware, kernel, operating system, and DNS application from vulnerabilities and attacks. Most of these controls are similar to or the same as those you've likely implemented for other infrastructure elements. Over the next five chapters, we will build on these foundational security outcomes and emphasize them with respect to respective DNS-specific vulnerabilities. In particular, we will review each major attack vector; define event detection approaches along with containment and mitigation measures, and data security and defensive strategies.

7

SERVICE DENIAL ATTACKS

INTRODUCTION

Denial of service (DoS) or distributed DoS (DDoS) attacks are brute force high volume data traffic attacks which are launched against a target to render services running on the target device unavailable to legitimate users of those services. These types of attacks have been launched at various types of network components over the years, including web servers, email servers, and DNS servers.

Beyond direct frontal service denial attacks, other forms of this attack generate volumes of traffic using a spoofed source IP address, matching the target's IP address; this generates voluminous response packets to the target's IP address. This reflector type of attack often leverages large numbers of queries to DNS servers using the source IP address of the intended victim.

We'll discuss these DNS-specific DoS attacks that exhaust DNS server resources using the DNS protocol or IP packets in general. We'll begin by reviewing the various attack forms, introduced in Chapter 4, then discuss detection and defensive approaches.

Denial of Service Attacks

DoS/DDoS attacks are invoked by an attacker with the intent to flood the DNS server with bogus DNS requests or other irrelevant packets, overwhelming its ability

DNS Security Management, First Edition. Michael Dooley and Timothy Rooney.
© 2017 by The Institute of Electrical and Electronic Engineers, Inc. Published 2017 by John Wiley & Sons, Inc.

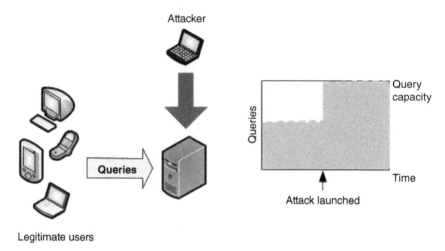

Figure 7.1. Denial of Service Attack

to process legitimate DNS queries. From the DNS server's perspective, it merely attempts to process each packet as it is received. As the volume of bogus packets is intensified beyond the query response rate supported by the server, the proportion of legitimate queries processed lessens and DNS resolution services capacity drops precipitously to only that small proportion that is processed. Ultimately, the attack may crash the server.

As illustrated in the crude query processing chart on the right side of Figure 7.1, prior to the attack, the server is receiving queries well within its capacity to respond. The attacker substantially increases the volume of queries in an effort to inundate the server and reduce or eliminate DNS resolution.

The types of DoS attacks against DNS may be in the form of the following:

- DNS query flood – the attacker issues a large number of DNS queries beyond which it has capacity to resolve.
- UDP packet flood – an attacker may issue large numbers of UDP packets using random UDP destination port numbers, forcing the server to respond with an ICMP Destination Unreachable message for each.
- TCP SYN attack – while DNS typically utilizes UDP, TCP is permissible, and the SYN attack involves the attacker opening a TCP connection by sending the TCP SYN message from varying source IP addresses and/or ports, then ignoring the SYN-ACK thereby not completing connection establishment with the third message of the three-way handshake. While awaiting each ACK, the server keeps the half-open TCP connection pending, and ultimately depletes its capacity for TCP connections.
- ICMP flood – the attacker issues a constant stream of ICMP packets to the server which uselessly occupies its processing capabilities.

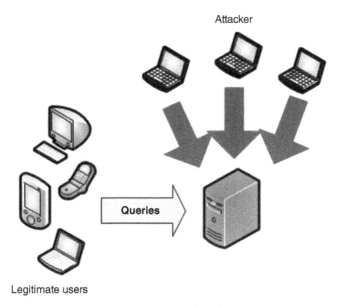

Attacker

Queries

Legitimate users

Figure 7.2. Distributed Denial of Service Attack

Distributed Denial of Service A variant of the DoS type of attack is the use of multiple distributed attack points and is referred to as a DDoS attack. The intent is the same, though the scale is larger, with multiple attack origination points.

Attackers can enlist others to manually conduct an attack on a target simultaneously as depicted in Figure 7.2. However, in many cases, the use of bots installed on other computers within the enterprise or on the Internet can be enlisted to join in the attack.

Bogus Domain Queries This attack attempts service denial through the flooding of a recursive server with queries for bogus domain names. This causes the server to initiate a "wild goose chase" and utilize resources to futilely locate an answer from the authoritative server within the domain tree as illustrated in Figure 7.3.

In addition to processing a high volume of such queries as in a typical DoS attack, the recursive server expends resources iterating queries to name servers within the domain tree in an attempt to identify the authoritative servers for each bogus domain. Ultimately, query errors will be returned for a lame delegation or NXDOMAIN responses but the sheer volume of such pending queries can inhibit its processing of legitimate queries.

Pseudorandom Subdomain Attacks

A variant of the generic bogus domain query attack focuses queries on one particular domain served by a set of authoritative servers. This attack vector has been

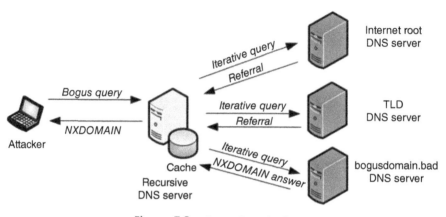

Figure 7.3.   Bogus Domain Query

shown to impact not only the servers authoritative for the targeted domain but recursive servers awaiting responses from these authoritative servers. This attack, called a pseudorandom subdomain (PRSD) attack features an attacker launching a large number of queries requesting resolution data for PRSDs beneath a target domain, let's say example.com. Thus an attacker queries for names like iopqewf.example.com, a84fj.example.com in large volumes. The large volume of queries can inundate the DNS servers authoritative for the example.com domain, thereby denying service.

Worse still, the ripple effect on the recursive server(s) to which the queries have been launched, for example, the attacker's ISP's DNS servers, can be debilitating. Once the authoritative servers have essentially crashed, the recursive server continues processing queries as portrayed in Figure 7.4. As the number of outstanding unanswered queries grows, the ability of the recursive servers to handle new legitimate queries diminishes, thereby reducing or even denying recursive DNS services for the ISP's customers.

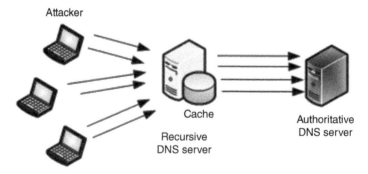

Figure 7.4.   PRSD Attack

Reflector Style Attacks

The reflector form of attack attempts to use one or more DNS servers to send massive amounts of data at a particular target, thereby denying service for the target machine. Accomplishing this type of attack relies on leveraging DNS servers (or routes to DNS servers) which do not perform ingress IP filtering and on DNS servers configured to enable recursion. Alternatively, this form of attack features an attacker querying "open resolvers" or Internet-facing DNS servers configured to enable query recursion. While recursion should be disabled for authoritative external DNS servers, unfortunately, there are millions of so-configured servers operating on the Internet today according to the Open Resolver Project (37).

Upon receiving a query from a given IP address, each server will perform its recursion function and respond accordingly to the purported requesting IP address as shown in Figure 7.5.

The following two major classes of this type of attack are defined which differ only in the query type issued to increase query answer size.

- Reflector attack – The attacker issues numerous queries to one or more DNS servers using the target machine's IP address as the source IP address in each DNS query. This attack could be issued using authoritative or recursive DNS servers which will happily respond accordingly to the source IP address. If several servers are queried at the same time, the volume of DNS response packets can become very large.
- Amplification – Using the reflector approach while querying for resource record types with large quantities of data such as NAPTR-, ANY-, and

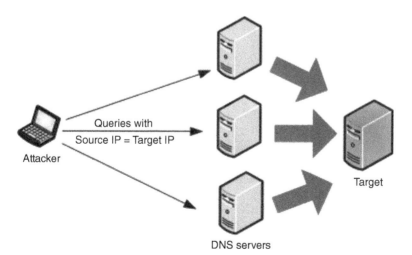

Figure 7.5.   Reflector Style Attack

DNSSEC-signed answers amplifies this attack by providing much larger response packets. Each responding server responds with the data to the "requestor" at the spoofed IP address to inundate this target with a large data flow, amplifying the attack volume with respect to typical query answers.

DETECTING SERVICE DENIAL ATTACKS

Symptoms of service denial attacks generally include a slowing or a complete halting of the ability to receive a response from the victim system. Your support desk may receive an influx of calls regarding the inability to reach a given server or that the network is "slow." A server of any type under a DoS attack will have its resources monopolized primarily in processing attack packets.

Server hardware components and processes should be monitored on an ongoing basis to measure server status and operational capacity. Monitoring of server CPU, memory, hard drive, and I/O utilization to capacity with thresholds on sustained operation at capacity in any one of these components should trigger an alert for investigation. Monitoring and alerting based on DNS daemon status and use of these resources provides a more granular window into the load on the DNS service running on the server as well as its load on the hardware platform.

Traffic meters measuring and tracking IP traffic levels on key segments of your network can provide a baseline or "normal" level of traffic when analyzed over time. A sudden dramatic rise in traffic could be an indicator of a form of DoS attack underway. Such an event is not always the result of an attack. An organization sponsored event, press release, or viral social media post could cause a rise in inbound traffic for legitimate if not desired users attempting to access your organization's resources. In this case, analysis of DNS query packets should indicate that a large proportion of the incremental traffic relates to the associated event or post.

Another potential cause of a sudden rise in traffic could be an outage elsewhere in your network where packet traffic has been rerouted, increasing the packet volume within the unaffected infrastructure. The incremental traffic volume should correlate to the magnitude of the outage, so the volume for a remote site outage versus a centralized datacenter should be commensurate with the surge in traffic.

DNS packet and query measurements as recorded at each server and tracked over time can likewise provide a baseline against which anomalous traffic patterns can be compared. Ideally, measurements at multiple layers of the stack, that is, for IP, UDP/TCP, and DNS layers provide the granularity to identify UDP floods versus DNS protocol layer attacks.

If the service denial attack uses the DNS protocol, it's beneficial to analyze the DNS traffic flows starting with the proportion of queries and answers. A large proportion of queries over answers could indicate the querier as a willing or unwitting participant in a DoS or PRSD attack, while a large proportion of answers over queries is a likely indicator that the answer recipient is the target of a reflector or amplification attack. Besides gauging the proportion of queries and responses, the presence of a larger than normal set of NXDOMAIN and SERVFAIL responses from your recursive server could indicate bogus queries or PRSD attack.

In summary, key service denial type attack indicators include

- Dramatic rise in IP packets measured on your network and/or at DNS servers
- Elevated memory or CPU utilization as measured on your DNS servers
- Traffic rise uncorrelated to any noteworthy announcement or marketing event
- Traffic rise uncorrelated to a network or DNS server outage elsewhere in the network
- Large rise in packets destined for a particular IP address
- Relatively small number of source IP addresses from which a large proportion of the traffic originates, for example, "top talkers"
- Disproportionate balance of query/answer transactions for DNS transactions in particular. Where the number of queries far exceeds the number of responses, bogus queries or PRSD attack may be indicated; if responses far exceed queries, this could signal a reflection or amplification attack
- Relatively large quantity of NXDOMAIN and SERVFAIL DNS responses (indicating potential bogus queries or PRSD attack)

DENIAL OF SERVICE PROTECTION

Effective DoS mitigation requires constraining the level of server processing for attack packets such that legitimate packets are still handled properly to minimize collateral damage. Methods to reduce the impact of various forms of service denial attacks are discussed in this section.

DoS/DDoS Mitigation

Upon detection of a DoS attack, the source IP address from which a high volume of packets is being received should be blocked or at minimum throttled by dropping a percentage of inbound packets. The lower in the stack these packets can be discarded, the fewer system resources will be consumed. For example, dropping packets at the server kernel layer obviates the need to DNS service processing in general.

But if you are suffering from a DDoS attack, blocking a single IP address will not sufficiently curtail the attack; if possible block or rate limit all of the participants in the attack if possible though these days with attackers enlisting thousands of bots in some attacks, the address blocking approach alone will not suffice and some of the following tactics may also be enacted.

- Inbound ACLs at the server level regarding from which IP networks or hosts the server will process queries (recursive servers).

- Inbound ACLs at the DNS service level to constrain from what IP hosts or networks the DNS service is permitted to process a DNS transaction of a given type; for example, query versus zone transfer.
- Inbound rate limiting of packets from specific sources and/or by type (TCP, UDP, DNS, etc.). Many DNS appliance vendors support configuration of rate limiting on each network interface. However, you can also configure rate limiting on Linux DNS servers using iptables to limit the number of incoming DNS queries from a given source IP address(es) and subnet(s). The `hashlimit` switch in particular enables you to limit for each host within a given CIDR-specified network.
- If you've enabled TCP for DNS on your recursive server, you can also use iptables to limit the number of inbound TCP connections.
- DNS anycast deployment where multiple DNS servers use a common IP address. This approach was proven effective against a DDoS attack against Internet DNS root servers in February 2007 (authoritative servers).
- Consider using a DNS hosting provider to supplement your in-house external DNS deployment (or vice versa, consider supporting your own in-house deployment to supplement your hosting provider). Or consider using multiple hosting providers. The October 2016 DDoS attack on the DynDNS infrastructure highlights the collateral damage impacting organizations that "solesourced" DNS hosting services from an attack on your hosting provider. Diversifying providers including your own server hosting can offer multiple paths to resolution and leave you less vulnerable to a DDoS attack on your service provider.
- Implement DNS cookies to enable lightweight message authentication of DNS requests and responses without requiring configuration.

DNS Cookies DNS cookies are specified in RFC 7873 (24) and define a lightweight DNS message authentication mechanism without pre-configuration. If both the client and server support DNS cookies, this scheme enables basic authentication of each party, thereby mitigating DoS attacks that utilize spoofed source IP addresses. DNS cookies do not protect against DoS attacks from correct (nonspoofed) IP addresses, though they can aid in quickly identifying the offending host for remediation.

When formulating a query, the client populates the EDNS0 "cookie" option with either a cached server cookie for the server to which the query will be directed or with a self-generated cookie comprised of a pseudorandom function of its IP address, the server IP address, and a random quantity. The server will reply a server cookie generated as a pseudorandom function of the client IP address, the client cookie, and a random quantity.

DNS cookies have not as yet been widely implemented, though it is natively supported in ISC BIND 9.11 and Knot Resolver 1.1. BIND uses queries with cookies to whitelist (i.e., process) such queries when configured for response rate limiting (RRL).

Bogus Queries Mitigation

Bogus query attacks attempt to consume recursive DNS server resources through the issuance of a large quantity of queries for random domains. The recursive server seeks resolution down the domain tree and ultimately no delegation exists for the lengthy domain name being queried. For example, queries in the form of <random label>.<valid subdomain(s)>.<valid domain>.<TLD>, where the random label and the valid [sub]domains may be altered. In the meantime, the server has several outstanding queries that it is chasing in an attempt to resolve. The ultimate result of NXDOMAIN serves as little consolation for all of the effort! Multiply this by thousands at nearly the same time and the recursive server resources may quickly become exhausted.

While standard port and protocol rate limiting tactics described above can help stem the flow of inbound DNS queries from a given host in general, a distributed bogus query attack could bypass this, especially given the fact that one inbound query results in multiple outbound queries and query state management cycles on the recursive server for non-cached entries. Implementing inbound query rate limits on your recursive servers can help to reduce the server load processing of a disproportionate number of queries.

The Unbound server provides a `jostle-timeout` option that enables the specification of a timeout value to use when the server is busy. This parameter is specified in units of milliseconds and the default is 200. As the number of arriving queries rises such that query timeouts rise above this value, the server will allow half of the outbound queries to run to completion but will apply the timeout value specified in this parameter to the other half.

The ISC BIND server supports a `fetches-per-server` option to limit the number of outbound pending queries for a given server. A second parameter for this option indicates how the server should respond to queries when the limit is exceeded: either drop the query or return SERVFAIL. Note that for BIND versions 9.9.8 and 9.10.3 and above, BIND must be built with the `-enable-fetchlimit` configure option, while in BIND 9.11 this feature is enabled by default. The query quota is dynamically sized based on the ratio of timeouts to successful resolutions. You can tune this dynamic sizing using the `fetch-quota-params` option which specifies four parameters.

- The number of queries between which the quota is recalculated; the default is every 100 queries.
- Threshold timeout ratio above which the server query load is considered acceptable and the quota may be raised; the default is 0.1 (10%).

- Threshold timeout ratio above which the server query load is excessive and the quota may be lowered; the default is 0.3.
- Weighting assigned to the most recent measurement period when averaging it with the previously measured timeout ratio; the default is 0.7.

PRSD Attack Mitigation

The PRSD attack is a variant of the bogus queries attack where the <valid domain> portion of each successive query is the same, for example, queries to fe9efas.example.com, oie8f9.example.com. The intended target of this attack is actually the DNS servers authoritative for the <valid domain>, example.com in our case. If the attacker employs one or more attacker resolvers behind a common recursive server, the recursive server may be victimized by the PRSD attack. While example.com likely resides in the recursive server cache, the random subdomain label likely does not, spurring a query to the authoritative servers for each. As the authoritative servers become overwhelmed and fail to respond, the recursive server's resources become consumed with tracking and awaiting responses to the outstanding queries.

The `jostle-timeout` and `fetches-per-server` options for Unbound and BIND, respectively, are helpful in throttling outbound queries to each authoritative server. BIND also provides a `fetches-per-zone` quota option that helps further contain queries for the victim domain. The single parameter setting for this option sets a hard limit on the number of outstanding queries permitted for the given domain including all subdomains.

BIND also provides a `max-clients-per-query` option which enables specification of the maximum number of simultaneous outstanding queries of the same name and type. The server issues one iterative query to obtain the authoritative response and responds to each client accordingly. However, in cases where the hostname, subdomain, or any label in the name varies, this option does not apply.

If these measures prove insufficient, even after tweaking numbers down, you may need to manually configure your recursive server as authoritative for the target domain to suspend recursion to that domain, example.com in this case. Or you can temporarily add the domain to your DNS firewall. In either case, do not forget to undo any such manual changes after attack recovery.

Reflector Mitigation

A reflector style attack features a series of DNS queries with a spoofed source IP address, to which the attack is targeted. The attacker attempts to launch a large quantity of such queries in order to flood the target at the spoofed source IP address. Such attacks where queries are issued with the intent of triggering responses much larger in size that the request, such as for DNSSEC-signed answers, amplify the traffic load on the target and are fittingly termed amplification attacks.

- Configuring IP address filtering using reverse path forwarding on your (and hopefully your ISP's) routers in accordance with BCP38 (39) can help reduce the success of spoofing.
- Most major DNS servers also support RRL features to help reduce the impact of such an attack. Generally, response rate limiting first requires detection of an attack through the observation of several successive nearly identical query responses destined to a common IP address.

RRL implementations typically utilize a dynamic response "credit bucket" scheme, whereby credits are allotted to each client, and as responses are delivered, response credits are reduced accordingly. If zero credits exist for a given client, the response is dropped. Credit status is reviewed periodically to refresh and renew credits.

NLnet RRL Parameters   NLnet's NSD server provides the following RRL configuration parameters definable within the server section of the nsd.conf file. The whitelist parameter is useful for defining popular domains for which a different RRL policy may be applied. The following parameters are definable in version 4.1.12; check the man page for your version for any variations.

- `rrl-size:` *numbuckets* – the number of client bucket entries in the form of hash table size; the default is 1 million. Queries from a given address block are analyzed within the same bucket. The size of each address block is defined by the `rrl-ipv4-prefix-length` and `rrl-ipv6-prefix-length` parameters for IPv4 and IPv6 addresses, respectively
- `rrl-ratelimit:` *qps* – the maximum queries per second (qps) allowed from one query source
- `rrl-slip:` *numpackets* – discard *numpackets* packets before sending a SLIP response (a response with the DNS header's truncated (TC) bit set to 1)
- `rrl-ipv4-prefix-length:` *length* – IPv4 prefix length for netblock address grouping, default is 24
- `rrl-ipv6-prefix-length:` *length* – IPv6 prefix length for netblock address grouping, default is 64
- `rrl-whitelist-ratelimit:` *qps* – the maximum qps from one whitelisted query source

ISC BIND Response Rate Limiting   The ISC BIND implementation enables configuration of a `rate-limit` stanza within the options or view block with several parameters including `responses-per-second` which sets the initial maximum number of permitted responses over a user-settable window of time after which the rate may be adjusted. As of BIND 9.11, the parameters configurable within the `rate-limit` block include the following. Please consult the BIND Administration Reference Manual for your BIND version.

- `responses-per-second` *number* – limit on nonempty responses for a valid domain name (qname) and resource record type (qtype)
- `referrals-per-second` *number* – limit on nonempty referrals for a valid qname and qtype
- `nodata-per-second` *number* – limit on NODATA responses for a valid qname regardless of qtype
- `nxdomains-per-second` *number* – limit on NXDOMAIN responses for a given qname
- `errors-per-second` *number* – limit on DNS errors other than NXDOMAIN (REFUSED, FORMERR, and SERVFAIL) for a given qname and qtype
- `all-per-second` *number* – limits UDP responses of any kind
- `window` *number* – window length in seconds over which response credits are tallied and averaged
- `log-only` *yes_or_no* – when yes, no queries are dropped but logging ensues – for testing purposes
- `qps-scale` *number* – used to dynamically scale rate limits; when the server qps exceeds this qps-scale parameter value, limits on responses, errors, NXDO-MAINS, and "all" are decreased by the ratio of the observed qps/qps scale
- `ipv4-prefix-length` *number* – size of IPv4 blocks within which all queries are treated as one bucket
- `ipv6-prefix-length` *number* – size of IPv46 blocks within which all queries are treated as one bucket
- `slip` *number* – discard *number* packets before sending a SLIP response (a response with the DNS header truncated bit set to 1)
- `exempt-clients` {*address_match_list*} – clients identified by the address_match_list are exempt from rate limiting
- `max-table-size` *number* – maximum size of the table used to track requests and apply rate limits
- `min-table-size` *number* – minimum size of the table used to track requests and apply rate limits

These parameters of both implementations enable granular control of rate limiting behavior, though you may need to experiment with tweaking various settings to arrive at the optimized behavior you desire. BIND 9.11 also added the `minimal-any` option to configure BIND to return a single RRSet arbitrarily selected instead of all matching RRSets as a measure to mitigate reflector attacks leveraging ANY queries.

Review server logs regarding dropped queries and automated rate manipulation by the server to identify the onset of an attack and to review attack mitigation during the recovery phase.

Knot DNS Response Rate Limiting   Knot DNS enables definition of a rate limit and slip interval within the server section, such as the following:

```
server:
    rate-limit: 100       # Allow 100 responses/sec for each flow
    rate-limit-table-size: 393241    #default hash table buckets
    rate-limit-slip: 2  # Every other response slips
```

Knot DNS also provides the ability to whitelist sets of IP addresses, subnets, and ranges for which to disable rate limiting with the `rate-limit-whitelist` parameter.

PowerDNS Parameters   PowerDNS enables you to configure a lua script to handle RRL such as those in email lists such as those published by Mark Scholten (45).

SUMMARY

This chapter discussed various forms of service denial attacks that can be brought to bear on your DNS servers and that can attempt to utilize your DNS servers as attack participants. Prepare for these attacks by applying defensive approaches we discussed in this chapter. Monitor for potential attack manifestations as early after the onset as possible, and observe the effectiveness of your defensive strategies. Additional controls such as manual reconfiguration or static rerouting may be required if defenses prove inadequate during the attack. After the attack has been mitigated and services restored, review lessons learned and apply them to updating defensive measures and settings.

8

CACHE POISONING DEFENSES

INTRODUCTION

A recursive DNS server performs the task of seeking the answer to a query received from a resolver, typically a stub resolver or another DNS server from which it forwards queries. Answers received during the resolution process across the set of clients querying the server are cached within the recursive server's memory. For this reason, recursive servers are also commonly known as caching servers. The cache enables the recursive server to seek to answer a query from what it already "knows" as accumulated in its cache. For a given query whose answer resides in cache, the server responds with the cached answer each time a query for this same information is received, for a time period until the record's TTL expires. Thus, the recursive server can respond immediately without initiating multiple iterative queries down the domain tree to the appropriate authoritative DNS servers.

Caching of this resolution data improves query performance, but this data cache also represents an attractive attack target as a means to steer not just one client, but several clients to an imposter website to gather personal or financial information. If an attacker can modify the cached record for the popular www.example-bank.com site, for example, to point to the attacker's website which looks like the bank's website, many users could fall victim to unwittingly sharing login credentials as well as account and personal information. All clients configured to query such a poisoned

DNS Security Management, First Edition. Michael Dooley and Timothy Rooney.

caching server would be vulnerable to this redirection while the forged record resides in cache.

ATTACK FORMS

To corrupt the cache, the DNS query response from the attacker must reach the server before the legitimate response and map to an outstanding query for which the recursive server is awaiting a response. The server will map a received answer to a previously issued query by matching the following fields in the response:

- The source IP address of the response maps to the destination IP address of the query and the destination IP address matches the address of this server.
- The destination port of the response with the source port of the query and the answer's source port is 53 (or more correctly matches the query's destination port with port 53 being the DNS well-known port).
- The DNS transaction ID within the DNS header matches on both the query and the response.
- The Qname, Qclass, and Qtype in the question section matches on both the query and the response.
- The domain names in the Answer, Authority, and Additional sections of the response must fall within the same domain branch as the Qname. This is known as the bailiwick check. Thus, a server should not cache information for example.com within the Additional section if the Qname was within the example-bank.com domain.

The attacker may send multiple responses with varying parameters, and the more responses he or she can send before the legitimate answer is received, the higher the probability of successful poisoning. A variety of cache poisoning attack vectors have been observed, each attempting to force a recursive server to accept and cache a falsified answer to an otherwise legitimate DNS query as we'll summarize next.

Packet Interception or Spoofing

Like other client/server applications, DNS is susceptible to "man-in-the-middle" attacks where an attacker responds to a DNS query with false or misleading additional information. In this scenario, the attacker has network and routing connectivity to the target recursive DNS server and spoofs the legitimate DNS server response, matching all fields accordingly, leading the recursive server to resolve and cache this information.

In order to successfully poison a targeted recursive server's cache, the attacker typically needs to send numerous responses within a brief time window, with one correctly matching all of the DNS answer criteria. And if the attacker fails to match these parameters before the legitimate response is received, the server won't issue another like query until the record's TTL expires, which could be hours or days. If the www.example-bank.com address record resides in cache, the cached answer will be returned immediately with no recursion.

If the attacker is actually able to see the query message as a true "man-in-the-middle" attack, the formulation of a response could be rather trivial. However, if an attacker has access to your Internet traffic, there is a much broader cause for concern. In the more likely case, where the attacker triggers the query, he or she then may send back numerous responses, varying the parameters on each response.

ID Guessing or Query Prediction

ID guessing refers to the attacker's ability to guess and match the DNS message transaction ID. The transaction ID field of the DNS packet header is 16 bits in length, as is the UDP port number field. Older versions of recursive DNS servers merely incremented the transaction ID field and selected a source UDP port number from a small pool of numbers. Even if the transaction ID is randomized, the degree of randomization depends on the entropy or randomness of the seed applied by the server. In systems with weak entropy, an attacker could possibly predict future values based on a small sample of past values.

The attacker's chances of success improve if the recursive server issues multiple query packets with the same question. This increases the pool of transaction IDs for which the server will accept a matching answer. And depending on the randomness of your server's transaction ID generation algorithm, the attacker may further improve his or her odds. Michal Zalewski published a white paper (46) about randomness and TCP sequence numbers which can be applied to randomness and DNS transaction IDs. He applied phase space analysis to characterize predictability based on various operating systems. A CERT vulnerability note (47) summarized the associated cache poisoning vulnerability and the increasing probability of an attacker correctly guessing with increasing packets sent by the recursive server.

Name Chaining

This attack features the provision of supplemental resolution information usually within the Additional or even Authority section of the DNS response packet, thereby poisoning the cache with malicious resolution information. Oftentimes an attacker-supplied resource record's domain would fall into a different branch of the domain tree.

The Kashpureff attack of 1997 utilized this form of attack in an attempt to subvert the authority of the Internet DNS root name and addressing authority of the time, the InterNIC. Eugene Kashpureff issued queries to recursive servers querying for his alternative NIC, alternic.net. These DNS queries worked down the domain tree to

Kashpureff's alternic.net authoritative name servers which responded in kind; however, they included forged internic.net IP addresses within the Additional section of the response. The recursive server cached not only the altnernic.net answer but also the internic.net answer, both of which pointed to Kashpureff's site. Future queries to internic.net were redirected to alternic.net.

Recursive servers which implemented the *bailiwick check* were not vulnerable to this attack. The bailiwick check, which all modern recursive servers implement, should inhibit these attempts to force the recursive server to cache this "irrelevant" domain information. This check negated attackers' ability to simply add poisoned records that fell outside of the domain branch of the query. Thus, attackers could no longer trigger arbitrary queries to domains they controlled to provide poisoned ancillary data, but would need to poison servers' cache by providing a falsified query answer to the targeted domain itself. Thus, to poison the cache for www.example-bank.com, the attacker would have to detect or initiate a query to this domain and provide an accepted (falsified) answer before the legitimate server.

The Kaminsky DNS Vulnerability

So far, we've seen that the attacker must issue a targeted query after the corresponding record's TTL has expired and before another user issues the same query without the attacker being ready. If the attacker is ready and observes or initiates the query for www.example-bank.com, the attacker can formulate a response with the answer section falsified with an address for their own web server. But they still need to match the other parameters from the query: IP addresses, UDP port numbers, DNS transaction ID, bailiwick, and question. Thus, while not impossible, it is still rather challenging to successfully poison a DNS server's cache. Or is it?

What if the attacker could initiate the attack by triggering a query for a resource record within the target domain at will? Assuming the attacker resides outside the confines of your network, an arbitrary query should be ignored if you've properly disabled recursion on your external servers and blocked inbound DNS queries. If the attacker is within your network, either directly or via a malware installation, your recursive servers are vulnerable. But your recursive servers are vulnerable even without malware infestation. If an attacker can phish users to their website, their web page may contain multiple links to trigger DNS queries to open the door for poisoning. In addition, most email servers issue DNS queries upon receipt of email to check spam filters configured as DNS records. And if you haven't configured reverse path filtering, the attacker could issue DNS queries directly using spoofed internal addresses (the public addresses of your Internet caching servers) in order to trigger external queries for records not in cache.

Dan Kaminsky, the Director of Penetration Testing at IOActive at the time identified a vulnerability which broadened the scope of cache poisoning attacks (48). Whereas typical cache poisoning attacks attempt to manipulate resource records within the answer section for a given destination, the Kaminsky attack seeks to manipulate cache via resource records populated within the Authority and/or Additional section(s), while remaining within the same bailiwick. This enables an attacker to repoint queries to its own name servers as authoritative for the target domain.

Let's say an attacker wishes to target a popular domain, say example.com. A "standard" cache poisoning attack may attempt to corrupt the cache for the www.example.com record. But if the attacker can essentially assume control of the example.com domain, he or she can provide falsified answers for the www record or any other record within the domain at will. However, if this record is frequently queried, the legitimate answer may already reside in cache and it may be difficult to predict when the cached record will expire, prompting a new outbound query which serves as an opportunity to attack.

Under the Kaminsky attack, the attacker initiates queries for random hostnames within the target domain, for example, 1.example.com, www83932.example.com. Such querying for domain names within the target bailiwick that are unlikely to already reside in cache should prompt the server to resolve these by querying other name servers. When such a query is initiated, the attacker may then start sending responses to the query each with varying destination UDP port numbers and transaction IDs in an attempt to provide a match to convince the recursive server of the seeming legitimacy of its response.

An attacker could initiate such queries by creating a web page with several tags with URLs such as img or link tags, causing the web browser to attempt to resolve each URL using DNS. If you view the source of most any web page, you'll see several href and src tags which point to domain names, each requiring resolution to fetch the corresponding image or resource. Thus, the attacker knows what queries will be asked, that is, the Qname, the set of valid name servers that will be queried within the target domain, so the challenge boils down to properly guessing the transaction ID and UDP port, the probability of which improves with multiple attempts as we discussed previously.

So we can initiate queries, but why do we want to poison a record that no one will query like 1.example.com? We don't, but here's where the second element comes in: we can provide a referral answer to the query for n.example.com with an NS record containing the target hostname, www.example.com. The corresponding glue record will consist of an A and/or AAAA record with the IP address of the attacker's web server.

Thus, we trigger a query to

- 1.example.com IN A

Our falsified answer comes back with an empty Answer section but

- `example.com IN NS www.example.com` (Authority)
- `www.example.com IN A` {attacker IP address} (Additional)

We still have to match the transaction ID, UDP ports, and IP addresses in our response. But we pass the bailiwick check and we can trigger multiple queries by incrementing or randomizing the hostname label which likely does not reside in cache. Eventually we will guess it correctly.

But this attack has further implications. The attacker's response could provide a query answer and provide his or her own name server names and IP addresses for the example.com domain. This would be conveyed within the Authority section of the response; that is, the NS and glue records for the example.com domain as illustrated in Figure 8.1. The attacker's answers meet the bailiwick requirement, so they would need to properly match the IP addresses, ports, and transaction ID to gain acceptance and poison the cache.

Figure 8.1 illustrates a contrast between what the recursive server may receive while under such an attack. A legitimate response to a query for the 1.example.com record is shown on the left of the figure, while a spoofed response to a query within the same bailiwick is shown on the right. The true legitimate response to a query for 1.example.com is likely NXDOMAIN, but if an attacker can respond first with matching IP, port, and TXID criteria, the answer will be cached, as would the name server and glue records from the Authority and Additional sections. And unfortunately, these poisoned NS and glue records enable an attacker to poison an entire domain branch. Future queries to hosts and subdomains of example.com would be directed to hack1.example.com or hack2.example.com while the TTL remains valid.

The power of this attack is that it not only enables the attacker to hijack an entire domain branch, it also equips the attacker with the questions to be asked and more importantly, when they will be asked.

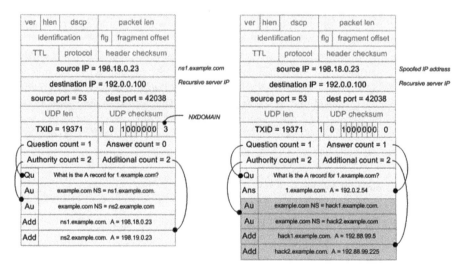

Figure 8.1. Legitimate versus Poisoned Response

In summary, Dan Kaminsky identified a major vulnerability that affected major DNS server vendors that enabled an attacker more control in attempting to poison a recursive server's cache by

- Allowing the attacker to initiate multiple DNS queries that likely do not reside in cache to trigger recursion
- Poisoning the cache of the target name and any other record within that domain by using the target name as a referral

If the attacker desired to poison more than one record but the entire example.com domain, they could substitute their own name server as the NS record in the response above. There is no requirement that an NS name for a given domain must reside within that domain; the bailiwick check does not apply. So the attacker could have answered with example.com IN NS ns.attacker-domain.com along with the corresponding glue. Clearly this form of domain hijacking could be devastating for the domain administrators and your end user clients.

CACHE POISONING DETECTION

The primary attack vector for poisoning DNS cache as we've seen consists of an attacker manipulating DNS responses by sending numerous answers to a given query with varying IP and DNS header parameters. A high volume of incoming DNS answers disproportionate to outgoing queries could indicate a cache poisoning attack. Of course, a DoS attack also features a large number of incoming packets, but cache poisoning attacks would generally feature bursts of similar responses in a more sporadic traffic flow.

If you are able to track historical answers for a given domain name and can identify changes in response data, this could be an indicator of a poisoning attempt; or not. The beauty of DNS from an administrator's perspective is that he or she can change IP addresses due to general renumbering, changes in ISPs, or other administrative reasons, and simply update the DNS to map the same domain name to a different IP address. Some changes in responses are to be expected. But if a high volume of incoming answers to a given question is detected, a verification step could be invoked to confirm the received answer.

Verification of responses by querying a second time could effectively highlight falsified data as it is unlikely the attacker would be able to successfully respond to the query. Of course, the query would need to be directed through a different recursive server since it has presumably already cached the suspect answer. Such a process is not a native DNS server function but requires a third party solution or manual process.

Upon detection of an attack, you should attempt to identify the source of the attack and if possible, block the sender. However, this may be difficult given the attacker's use of spoofed IP addresses. Meanwhile, dump your recursive DNS server's cache for forensic analysis later, then dump the cache to prevent further proliferation of the poisoned resolution data. If you're using BIND, you can use the `dumpdb -cache` remote name daemon controls (rndc) command to dump the cache and the `flushname`, `flushtree`, or `flush` rndc commands to flush the cache of a given domain name, the domain name and its subdomains, or the entire cache, respectively.

For Unbound, use `dump_cache` to dump the cache and `flush` with a zone name to flush the cache pertaining to the given zone for types A, AAAA, NS, SOA, CNAME, DNAME, MX, PTR, SRV, and NAPTR. Use `flush_zone` to flush the cache for the specified domain and its subdomains. For PowerDNS, use `dump-cache` to save the cache to a specified file name and `wipe-cache` to clear the cache for information related to specified domains in the wipe-cache command. For Knot DNS, the `cache.clear` command clears the cache, though you can also specify a domain to clear only cache entries relevant to the specified domain.

You can review query logs to determine which stub resolver devices had received the falsified response; then instruct users of those devices to flush their cache, for example, close and restart their browsers. Contact the DNS administrator for the domain for which the attacker poisoned your cache. They need to know of the attack on their namespace so they can devise a plan to secure it, namely by signing their zones.

CACHE POISONING DEFENSE MECHANISMS

First, please make sure your external DNS servers do not allow recursion. Your DNS servers receiving queries for records for which your servers are not authoritative should drop them immediately. Also, apply spam filters if you haven't already to block phishing emails from entering your networks. Curious users clicking a link to a web page could unwittingly trigger a cache poisoning attack by triggering attacker-controlled DNS lookups as we've described. Antivirus and anti-malware defenses for end user devices also can help reduce the incidence of known viruses and malware from instigating a cache poisoning attack against your recursive servers.

UDP Port Randomization

Beyond these preparatory steps, to protect your recursive servers, the simplest mitigation entails randomizing not only the transaction ID field but also the UDP source port. All recent releases of major DNS server vendors perform this automatically, so make sure yours do too. As we discussed, the 16-bit transaction ID field provides only 65,536 possible values. By randomizing the UDP port field, we add 11 bits (beyond the well-known port numbers), yielding $16 + 11 = 27$ bits or $2^{27} = 134$ million possible values. The probability of matching on both of these values is far lower than that of guessing the transaction ID alone.

For example, if an attacker can send 1000 responses prior to the legitimate answer arriving, they have only a 0.37% probability of guessing both fields correctly, whereas they'd have a 99.95% probability of guessing the transaction ID value alone. This assumes a very high degree of entropy and as the degree of true randomness diminishes, the higher the attacker's probability to successfully guess. Nevertheless, this strategy provides a vast improvement in protection from cache poisoning though it does not eliminate it.

Query Name Case Randomization

Some servers support the ability to introduce mixed case characters within a query, for example, wWW.exAmPLe.cOm, whereby the recursive server requires an exact case match on a valid answer as well. This technique requires the authoritative DNS server to replay back the question in the case presented, that is, instead of converting to lowercase. When case is honored, this added level of entropy inhibits successful guesses or at least makes it more difficult. Effectiveness of case randomization depends on the length of the Qname and the degree of randomness in selecting case.

DNS Security Extensions

The definitive solution to cache poisoning attacks requires the authentication of each DNS query answer as having been published by the authoritative domain owner. DNS security extensions (DNSSEC) specifications provide for query answer authentication as well as data integrity validation to assure no manipulation of query answers en route to the validating resolver. DNSSEC also provides authentication of information that does not exist in the domain's zone, or "authenticated denial of existence." DNSSEC provides these features by digitally signing DNS data.

Digital Signatures Digital signatures enable the originator of a given set of data to sign the data using a private key and a recipient to validate the data using a corresponding public key. DNSSEC uses asymmetric key cryptography, where a private/public key pair is mathematically bonded. In such a model, data signed with a private key can be validated by deciphering the data with the corresponding public key and vice versa, thus providing a means for holders of the public key to verify that data was signed using the corresponding private key. This provides authentication that the verified data was indeed signed with the private key. Digital signatures also enable verification that the data received match the data published and were not tampered with in transit.

Refer to Figure 8.2. The data originator, shown on the left of the figure, generates a private key/public key pair and utilizes the private key to sign the data. The first step in signing the data is to produce a hash of the data, also referred to as a digest. Hashes

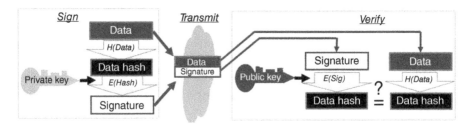

Figure 8.2.   Digital Signature Creation and Verification Process

are one-way functions* that scramble data into a fixed length string for simpler manipulation, and represent a "fingerprint" of the data. This means that it is very unlikely that another data input could produce the same hash value. Thus, hashes are often used as checksums but don't provide any origin authentication (anyone knowing the hash algorithm can simply hash arbitrary data). Currently supported hash and signature algorithms for DNSSEC include Diffie–Hellman, DSA/SHA1, DSA-NSEC3-SHA1, RSA/SHA-1, RSASHA1-NSEC3-SHA1, RSA/SHA-256, RSA/SHA-512, ECC-GOST, ECDSA Curve P-256 with SHA256, and ECDSA Curve P-384 with SHA384. The encryption algorithm transforms the hash with the private key to produce the signature.

Both the data and its associated signature are transmitted to or otherwise obtained by the recipient. Note that the data itself is not encrypted, merely signed. The recipient must have access to the public key that corresponds to the private key used to sign the data. In some cases, a secure (trusted) public key distribution system such as a public key infrastructure (PKI) is used to make public keys available. In the case of DNSSEC, public keys are published within DNS and signatures are included with query answers (i.e., the signed data).

The recipient computes a hash of the received resource record data, as did the data originator. The recipient applies the encryption algorithm to the received signature with the originator's public key. This operation is the inverse of the signature production process and produces the original data hash as its output. The result of this decryption, the original data hash, is compared with the recipient's computed hash of the data. If they match, the data has not been modified and the private key holder signed the data. If the private key holder can be trusted, the data can be considered secure.

When a resolver issues a query, the answer may contain multiple resource records. Many servers are often deployed for high availability or performance, so

---

* A one-way function means that the original data is not uniquely derivable from the hash. That is, one can apply an algorithm to create the hash, but there is no inverse algorithm to perform on the hash to arrive at the original data.

many address records may be returned in answer to a query for a given name. The set of resource records within the answer is referred to a resource record set (RRSet), which comprises one or more resource records with the same owner/name, RRType, and class. DNSSEC provides for the signing of the RRSet which provides a signature covering the set of answer data in the response. Any attempt to spoof or otherwise modify the RRSet data en route to the destination would be detectable by the recipient, that is, the resolver or more typically, its recursive/caching DNS server on its behalf. This feature makes DNSSEC an effective mitigation strategy against man-in-the-middle and cache poisoning attacks.

DNSSEC Trust Model   But just because an RRSet is signed, does that make it secure? Certainly not, as an attacker can define a falsified RRSet sign it with an arbitrary private key and willingly supply the corresponding public key to verify the signature. The digital signature standard (49) requires a mechanism to not only verify signatures but to confirm assurances that the signatures correspond to the intended signatory.

The DNSSEC specifications do not account for a secure key distribution system, so one or more trusted keys have to be manually configured on the resolver or recursive name server.* A trusted key is the public key authorized and published by the domain administrator for the given DNS zone. For each signed query answer, the resolver must not only validate the signature but must confirm that the public key used for validation corresponds to a trusted key for that zone.

As you consider how many different zones your resolvers query on a daily basis, you may be overwhelmed by the thought of configuring thousands of trusted keys, one per signed zone. Fortunately, DNSSEC implements a chain of trust that follows the standard DNS domain tree to enable validation of keys up the tree to the trusted root zone. Each domain from the root down to the zone signatory must be linked within the chain of trust to successfully validate signatory assurances. A break in the chain of trust invalidates the trustworthiness of the key unless the zone's or linked signed ancestor zone's public key is configured as a trusted key.

Configured trusted keys are referred to as trust anchors, and the root zone key may suffice as your single trust anchor or you may need to configure additional trust anchors for signed zones whose parent zones are not signed. We'll discuss how this chain of trust is linked with the domain tree in Chapter 9 when we discuss zone signing, but for now, suffice it to say that the chain of trust to a signed root zone simplifies the recursive server configuration for DNSSEC validation.

Just as passwords should be changed periodically, signing keys should be changed periodically. We'll talk the specifics regarding the mechanics for changing (rolling) keys in Chapter 9, but the impact on recursive servers is that if the root

---

* Pragmatically, the term "resolver" in the context of DNSSEC refers to the resolver function of the recursive server, which resolves the queried information and verifies signatures as well.

zone key changes without updating the trust anchor configuration, DNSSEC validations will fail. Fortunately, RFC 5011 (50) defines a means for recursive servers to
automatically detect trust anchor changes and to update corresponding trust anchor
configuration. RFC 5011 defines the procedures for authenticating new and revoked
trusted keys based on a manually configured initial trusted key. This "initial key"
serves as the "initial condition" in rolling forward over time with new, revoked, and
deleted root zone keys.

Configuring DNSSEC Trust Anchors   BIND, Knot, and Unbound support
automated trust anchor management as specified in RFC 5011. Thus, configuration
of the initial key comprises the bulk of work necessary to enable DNSSEC validation
for these reference implementations. Use secure network time protocol (NTP) to set
the server's time and date because DNSSEC signatures utilize absolute representation
of signature validity times (since and until) and not relative times.

BIND CONFIGURATION   For all currently supported BIND versions, configure
the initial trust anchor with the managed-keys statement within the named.conf file,
or just use the default provided with the BIND distribution in the bind.keys file.

```
managed-keys {
. initial-key 257 3 8
"AwEAAagAIKlVZrpC6Ia7gEzahOR+9W29euxhJhVVLOyQbSEWOO8gcCjFFVQUTf6v58fL
jwBd0YI0EzrAcQqBGCzh/RStIoO8g0NfnfL2MTJRkxoXbfDaUeVPQuYEhg37NZWAJQ9Vn
MVDxP/VHL496M/QZxkjf5/Efucp2gaD0X6RS6CXpoY68LsvPVjR0ZSwzz1apAzvN9dlzE
heX7ICJBBtuA6G3LQpzW5hOA2hzCTMjJPJ8LbqF6dsV6DoBQzgul0sGIcGOYl7OyQdXfZ
57relSQageu+ipAdTTJ25AsRTAoub8ONGcLmqrAmRLKBP1dfwhYB4N7knNnulqQxA+Uk1
ihz0=";
};
```

Next   configure   `dnssec-enable yes;`   and   `dnnsec-validation auto;`
within the options block of the server's named.conf file. Given the default
configuration of the managed-keys and dnssec-enable statements, the only thing you
really need to do is set dnssec validation to auto.

If you need to configure a nonroot trust anchor or if you require an internal root
server to enable resolution of internal domains only (e.g., for factory applications),
you can configure the trust anchor by setting `dnssec-validation yes;` and using
the trusted-keys statement which follows the same syntax of the managed-keys statement though without the initial-key phrase. Define your trust anchor key within the
key section of the trusted-keys statement.

```
trusted-keys {
. 257 3 8 "IlUvlbE4eVr23g0dS ... aPR1l8too30=";
};
```

For more details about further configuration details and troubleshooting, please consult the informative BIND DNSSEC Guide (51).

UNBOUND CONFIGURATION   Obtain the initial root zone key using the unbound-anchor command (52) though NLnet advises you to confirm trust. The trust anchor is stored by default in /usr/local/etc/unbound/root.key though you can move this using -a switch on the unbound-anchor command. Unbound must be able to read and write this file. In fact, the file should be world readable and unbound writable only.

Next configure the Unbound server to use this file as the trust anchor and to automatically update it based on root zone key changes.

```
server:
auto-trust-anchor-file:
"/usr/local/etc/unbound/root.key"
```

You can alternatively define a manual trust anchor, for example, for an internal root using either the trust-anchor-file or trust-anchor statements within the server block. For further details regarding configuration and testing, refer to the Unbound documentation (53).

POWERDNS CONFIGURATION   The PowerDNS Recursor does not support RFC 5011 as of version 4.0.1. Therefore, you need to configure the Recursor with your trust anchors and manually update them upon key roll. Enter your trust anchor(s) within the lua-config-file using the addDS function using input parameters of the zone to which the trust anchor applies, that is, "." for the root, and the DS RData for the root's public key. Enable DNSSEC validation using the dnssec parameter in the recursor.conf file. You can use dnssec process to have the Recursor perform DNSSEC validation if the AD or DO bits are set in the query; use dnssec validate to validate all queries.

KNOT RECURSOR CONFIGURATION   The Knot Recursor enables manual or automated trust anchor configuration. Use the -k switch with the kresd command to trigger automated root trust anchor lookup and storage in a specified filename.

```
$ kresd -k root.keys
```

Alternatively, you can manually configure trust anchors within the configuration file with the trust_anchors.file = 'root.keys' statement.

Lookaside Validation   DNSSEC Lookaside Validation (DLV) was devised before the root zone was signed to provide an "alternate" upstream chain of trust. DLV utilizes a centralized registry of signed zone public keys. By configuring the

DLV registry as a trust anchor, you thereby trust the DLV registry and all "child" zones which it authenticates.

These zones are not actual child zones of the DLV, but are zones that the DLV registry authenticates. Zone administrators may register their signed zone keys with the DLV registry in a secure manner to maintain this "lookaside" or "sideways" chain of trust, as opposed to the standard "upward" domain tree parent–child chain of trust. With the signing of the root zone in 2010 and most TLDs since then, the attractiveness of managing DLV linkages has vastly diminished, so we recommend you use the root trust anchor instead of DLV. The ISC DLV service which provided DLV registry services is scheduled to be disabled in 2017.

Negative Trust Anchors   As DNS administrators first implement DNSSEC, they sometimes make errors in properly aligning the keys, zone signing, timing, and linking the chain of trust. After all, public key cryptography is not natively an area of expertise for DNS administrators. Nevertheless, such misconfigurations can cause errors in DNSSEC validation, breaking the chain of trust.

If you detect validation failures for a certain domain, contact the domain administrator to convey the issue and to confirm the issue is indeed an error condition and not an attack such as a key compromise. If confirmed to be user error, you can define the domain as a negative trust anchor, which temporarily suspends the domain from requiring DNSSEC validation for successful resolution. Your users will be able to resolve DNS information for this domain, albeit without DNSSEC validation.

Ask the domain administrator when they expect to repair the DNSSEC configuration. You can obtain the domain administrator's email address from the zone's SOA record or possibly from whois. Try to perform a DNSSEC query for a record in the domain at the appointed time. If the issue persists, notify the domain administrator. No one likes to be a nag but if this domain is relatively popular, nagging may be the only means at your disposal to encourage resolution of the issue. Once resolved, you should remove the negative trust anchor to re-engage DNSSEC validation for the domain.

ISC BIND enables the setting of negative trust anchors via its rndc control channel. Use the `rndc nta` command to configure the specified domain as a negative trust anchor. You can configure the `nta-lifetime` and `nta-recheck` options within the named.conf options block to configure the default time until the negative trust anchor expires and DNSSEC validation is restored for the domain and the interval between automated DNSSEC enabled queries to the domain to detect resumption of proper DNSSEC validation, respectively. These options help automate the detection of resumption of zone signing without the necessity of nagging the zone administrator.

With Unbound, you can configure negative trust anchors using the `domain-insecure` directive within the `server` block, or using the unbound-controls command line with the `insecure-add` command. Unbound does not currently automatically detect restoration of zone signing. You can remove the domain-insecure statement for the respective zone or use `insecure-remove` from the command line to re-enable DNSSEC validation.

If you're using PowerDNS, you can configure the lua-config-file using `addNTA` with the domain name and optional comment, for example, `addNTA('example.com.', "Bad delegation pending repair")`. You can alternatively add, display, and clear negative trust anchors from the command line using `add-nta`, `show-ntas`, and `clear-nta`, respectively.

For the Knot Recursor use the `trust_anchor.set_insecure (domain list)` statement in the configuration file.

DNSSEC Deployment   In order to authenticate each DNS query answer, the answer must be signed. Thus, the authoritative server must publish signed zones and the recursive server must perform DNSSEC validation. Unfortunately, while 88% of US government zones are signed, only 8% of university zones and 2% of industry zones are signed according to NIST estimates (54). Thus, the majority of non-US government zones remain unsigned as of today. In terms of DNSSEC validating resolvers, APNIC measurements indicate about 15% of DNS resolvers support DNSSEC validation (55).

We've seen how simple it is to configure DNSSEC validation and we recommend you configure your recursive servers accordingly. However, given the modest deployment of signed zones, the additional measures of randomized transaction IDs and UDP port numbers must be implemented to protect your DNS caches. Case randomization may also be used for added entropy if your server supports this.

Last Mile Protection

Since the recursive server typically performs DNSSEC validation and not the stub resolver, resolution data are only validated between the recursive server and the authoritative server. Typically, there is an implicit trust network between the stub resolver and the recursive server. However, this may not be the case and this link between the stub resolver and the recursive server, sometimes referred to as "the last mile," may not be secured. We discuss here some approaches to secure this link should it be of sufficient risk to your organization.

DNS Cookies   We introduced DNS cookies in Chapter 7 as a means to mitigate DDoS attacks. DNS cookies provide a lightweight authentication mechanism for DNS clients and servers without requiring configuration a priori. The client creates a cookie by hashing its IP address, server IP address, and a random quantity and passes this to the server. The server can respond with its own cookie which the client would use in subsequent queries, based on a hash of the querier's IP address, cookie, and random quantity.

DNS over TLS   The use of DNS cookies could be used as a weak form of authentication. Alternatively, Microsoft requires an SSL connection from Windows clients to Windows DNS servers to secure this link. Other forms of authentication and encryption could be used as well. DNS over Transaction Layer Security (TLS) offers encryption of DNS queries and responses for resolvers and recursive servers enabling TLS support. TLS can be used for DNS queries and responses over TCP, while DNS over UDP transactions can be secured using Datagram TLS (DTLS). DTLS does impose packet size restrictions however, so transactions with large payloads such as DNSSEC responses may require fragmentation.

DNSCrypt   DNSCrypt is a protocol that provides security between DNS clients (resolvers) and the DNS recursive servers by applying cryptographic signatures to DNS messages to authenticate that the messages between the DNS resolver and the server are authentic. DNSCrypt is open source and there are several reference implementations that work on a variety of operating systems and devices including Windows, OSX, Linux, Unix, Android, iOS, and several versions of routers.

The DNSCrypt Client is installed on the DNS client system or device, which then communicates with the DNSCrypt Server that runs on the recursive DNS server. The DNSCrypt components operate as proxies. The client is configured to use the DNSCrypt Client proxy by configuring the client's resolver to point to localhost or to whatever IP address and port the DNSCrypt Client has been configured to run. The DNSCrypt Client is configured to sign and forward the queries to the DNSCrypt Server. The DNSCrypt Server validates the message signature and forwards it to a trusted DNS Resolver, typically running on the recursive server. Query answers are conversely signed and sent back to the DNSCrypt Client which verifies the responses and passes them back to the client resolver if genuine.

There are currently several public DNS Resolvers that host the DNSCrypt capability. A list of these servers can be found on links on the https://dnscrypt.org/ web site. You can implement just the DNSCrypt Client locally and point to one of these servers to use this protocol, or you can host your own DNSCrypt-enabled DNS server. The DNSCrypt protocol only allows for DNS traffic to be authenticated, but does not prevent DNS eavesdropping as DNS traffic is not encrypted. To protect against DNS eavesdropping, you would need to look at alternative technologies such as a Virtual Private Networks, which will encrypt the traffic.

9

SECURING AUTHORITATIVE DNS DATA

INTRODUCTION

This chapter describes the security of the DNS information you desire to publish to enable resolvers from within your organization, from partner VPN links, or from the Internet at large to resolve. As the administrator of your DNS zone information, it's your responsibility to publish host and IP address information (and other information) that your constituencies require to access services and applications they require and you support. You may want certain audiences to have access to a wider range of resolution data than others.

We discussed this in Chapter 5 with the deployment of trust sectors for major resolver audiences, namely internal, external, or extranet trust sectors. This basic partitioning should streamline the configuration of resolution data for each audience and minimize leakage, though you may require finer granularity particularly to segment internal audiences.

If you do require further granularity, for example, to publish resources containing confidential information for senior management, you could deploy an independent set of authoritative servers, establishing an additional trust sector as we discussed in Chapter 5.

DNS Security Management, First Edition. Michael Dooley and Timothy Rooney.
© 2017 by The Institute of Electrical and Electronic Engineers, Inc. Published 2017 by John Wiley & Sons, Inc.

ATTACK FORMS

The main objective of attackers attempting to manipulate authoritative data include modification of resolution data. With such power, the attacker can essentially make the domain disappear or he/she could repoint resolution data such that traffic is steered toward attacker resources for personal or financial data collection. Manipulation of resolution data can be effected by modifying data at rest on the authoritative server itelf or by falsifying the answer to a given query for the zone for which the server is authoritative as a cache poisoning attack. The latter form does not infiltrate the authoritative server itself but "impersonates" the authoritative server by providing a seemingly valid though falsified response to the querying recursive server.

Resolution Data at Rest

In the first scenario of an attacker manipulating the authoritative data at rest, an attacker is able to modify the resource record information published in your zone files. They can accomplish this by hacking the DNS server and editing the DNS configuration or zone files. This attack vector can be mitigated through host controls as we discussed in Chapter 6.

Another more subtle approach to manipulating published DNS information features the use of the DNS protocol itself using DNS Update messages to add to or modify resource record information in the target zone. An attacker may attempt to modify the zone information by sending DNS Updates to your authoritative master DNS server to perhaps modify, say your www resource record. This type of "attack" may alternatively be inadvertently initiated by a DHCP server which updates the DNS server with a host domain name and IP address mapping. Such an errorneous update may be innocent, but it can have detrimental impacts on the integrity of your namespace and therefore represents a risk that must be addressed.

A variation of the use of DNS Update messages to manipulate zone information instead mimics the mechanism you use to manage and configure your DNS servers. You may update DNS configurations via file editing (which would require host access and therefore can be mitigated with host controls), file transfer (e.g., from an IP address management (IPAM) system), or other technology your server may support such as rsynch.

Domain Registries

Another vulnerability to resolvers' ability to locate your domain information relates to the integrity of your parent and any other ancestor domains. An attacker modifying your name server information effectively hijacks your domain by changing the pointers down the domain tree to the attacker's name servers instead of yours. Typically, an organization will register a second level domain with a domain registrar for the desired top level domain (TLD). For example, one could register example.com with a .com registrar.

The domain registrar is authorized by the domain registry to uniquely assign subdomains (for .com in this case). Said another way, the domain registry technically manages the domain and is responsible for the data within and access policies for the repository containing domain registrations for immediate child domains. The registry publishes the domain zone file. A domain registrar is accredited by the corresponding registry and manages reservations of domain names within the registry repository.

There are over 1500 TLDs available to choose from, given the recent "new gTLD" program conducted by the Internet Corporation for Assigned Names and Numbers (ICANN). You can view the current list of TLDs, including the country code TLDs, internationalized TLDs, and new gTLDs on the IANA website (56). Contact a registry for the TLD for which you are interested in registering.

Consider the security practices of your domain registrar including the authentication and authorization technologies used for updating domain information and the use of DNSSEC. Registrars that lock domains require you to access your account to unlock the domain, then submit the update using an Authorization Code, then relock the domain. This process is more secure than merely submitting a change via email with just an authentication code; an attacker that can glean your authentication code, for example, via hacking your email account, could submit seemingly authorized changes (be sure to secure your email account as well). Such a two-factor change authorization process at the registry level is also critical to securing updates to your domain information. Such authorization assumes a secure account access process as well, including a notification process should the registrar detect a breach or better yet, whenever a change is made.

All new gTLDs support DNSSEC as mandated by ICANN and most other TLDs do as well, with 90% of all TLDs presently being DNSSEC-signed (57). You'll need to understand the DNSSEC update process as well, as your registrar will need to update your downstream pointing DNSSEC records (Delegation Signer, DS) when you roll your key signing key (KSK). We'll discuss this process later in the chapter.

One more thing to consider is whether you'd like your personal contact information published in the globally accessible whois database. This database enables people to query information for a given domain, including contact name, address, email, and telephone number. Most registrars offer an option to anonymize this information by using a "hostmaster" or similar generic label for the contact name and the registrar's address, email, and telephone number. This may be of particular interest if you operate a multinational network amid various regulations relating to the international transfer of personal information.

DNS Hosting Providers

If you publish your external DNS namespace utilizing a third party DNS provider instead of or in addition to operating external DNS servers in-house, you should consider similar precautions as you should with your domain registrar. Verify appropriate security practices for your provider including authentication and authorization methods, transaction encryption, and support of DNSSEC to secure your namespace's integrity.

Unlike registry or registrar compromise, an attacker with access to your external namespace can perform more surgical modifications. While an attacker can still resteer your domain (and therefore subdomains) to his or her own servers, he or she can also simply update a record or two such as your www or mail records to divert certain transactions. In fact the attacker could merely add their records to your existing RRSet in an attempt to siphon off a portion of your traffic while attempting to avoid detection through a drop in resolution traffic.

In such cases, monitoring traffic by RRType may be useful but drops within an expected variance would not likely trigger alarm. A better approach to detect such an attack is to simply confirm your zone information contains only that which you intend. You can detect zone changes through notification from your service provider and/or by retreiving the zone periodically and scanning it for changes to determine if the changes were authorized. This "diff" approach requires fairly regular polling since this will define your window of exposure before detection.

Just querying the serial number for each zone is certainly a simpler approach, but if the attacker can modify the zone with being forced to update the serial number, this method may not detect such an attack.

DNS Data in Motion

The cache poisoning attack to manipulate what others see as your authoritative data, as we discussed in the prior chapter, is best mitigated through the signing of your namespace via DNSSEC. In fact DNSSEC is the only means by which an authoritative zone publisher can mitigate poisoning of domain information communicated during the resolution process. And of course this mitigation is only applicable if the resolver validates DNSSEC signatures.

ATTACK DETECTION

Authoritative Data

You can install a file checker utility on your DNS servers to periodically produce DNS configuration and zone file signatures to detect signature differences of a given file to detect changes. Of course depending on the file, particularly a zone file, it may change quite frequently, so this detection method may be more annoying than useful. But for configuration and relatively static zone files, it may be a simple periodic check to verify file integrity. You can monitor DNS logs for zone changes as we'll discuss in the next section.

You can also periodically issue queries to your authoritative servers to confirm resolution data received matches what you expected. This would require a repository of expected resolution data separate from your DNS servers for comparison of query results. Of course any authorized changes made to resolution data must be reflected in the "expected data" repository as well.

Domain Registry

If your domain information related to your name servers is compromised, you will notice a precipitous drop in inbound query traffic. Recursive servers seeking your domain from the root servers down the domain tree will be pointed elsewhere thanks to the attacker's change of your NS, glue, and DS information in your parent domain. You should login to your registrar account to view the current configuration. If it has indeed been modified, correct the information, change your password, and notify the registrar of the breach.

Any falsified resource records provided to queries at your expense for your domain information will have been served by the attacker's servers and will be cached by recursive servers for the duration of the TTL, which could generally be up to a week. You will need to contact your subscribers or customers to inform them of the issue and how to rectify the situation by flushing relevant cache information.

Domain Hosting

Depending on the form of the attack, you may see similar symptoms as with a registry or registrar compromise if the attacker modifies or deletes zone information. If you see a vacuum of DNS traffic, you should access your domain account to verify. If the attacker changed your credentials, while you work with your domain provider to reauthenticate yourself to regain access, you can issue queries for records you had configured in the zone. This assumes you have a record or a copy of your external zone. In this regard, you may want to setup a DNS server as a slave to your service provider DNS servers, so you can receive zone level notifications of changes as they occur unless the attacker is clever enough to change this too.

Validating your zone contents through zone transfers if you have a valid server on the "allow-transfer" list of the master enables a straightforward means of viewing how the world views your zone. Short of this, issuing queries periodically for resolution data you've published rather frequently and alerting upon detection of changes provides another method to detect changes. You can ignore authorized changes of course but this can supply warning of an unauthorized change.

If your domain account has been compromised, notify your service provider and reset your credentials immediately.

Falsified Resolution

As the administrator of your authoritative DNS servers, you will not necessarily have visibility to cache poisoning attacks on recursive servers attempting to resolve your namespace. Cache poisoning attacks attempt to provide a falsified response for a given query before the authoritative server responds. From the authoritative perspective, your server receives a query and issues an answer but you have no visibility to other responses hitting the recursive server assuming they originate outside the bounds of your network. By signing your authoritative namespace using DNSSEC, you can assure validating resolvers of the integrity of your namespace.

DEFENSE MECHANISMS

Defending DNS Data at Rest

DNS server host controls must be implemented to protect your servers from unauthorized administrators or attackers from inadvertent or malicious access. Please refer to Chapter 6 for a discussion of relevant controls. Beyond protecting illicit access to the DNS server itself, for example, by obtaining the server command prompt locally or remotely, controls must be dispatched to protect against illicit manipulation of DNS-specific data.

Consider all of the methods that your DNS server implementation permits to enable changes to its configuration. If you manually edit files on the server, the host controls we mentioned above should cover most concerns. However, you should also go beyond server access to consider administrator permission assignments to enable access to the directories and files containing DNS configuration and zone data. If you're supporting other applications on the server with multiple administrators managing all or some of them, file system permissions grow in necessity.

Defenses against domain registry, registrar, or hosting provider attacks lie predominantly on the respective provider of these corresponding services. Proactive periodic checks on what your namespace looks like on the Internet is important for detecting unusual conditions that could indicate breach of integrity.

If you use an IPAM system to manage your DNS configuration, consider the communications channel between the IPAM system and each DNS server. Secure this connection using an authenticated, encrypted connection such as Transaction Layer Security (TLS), Secure File Transfer Protocol (SFTP), Secure Copy (scp), or other standard or proprietary protocol. Add the corresponding daemon(s) to your watch list of software for which you monitor vulnerabilities and patches.

If you enable your control channel on your DNS server to perform configuration or zone updates, be sure to secure this channel. BIND enables you to define a listen-on IP address and port as well as allow and deny access controls for address match lists (containing IP addresses, subnets, TSIG keys, and other match lists) within the `controls` statement. For NSD, run the `nsd-control-setup` command to generate TLS keys (self-signed certificates) to secure control channel transactions. You can configure the control IP address and port as well as certificate and key files within the nsd.conf file using the `control-enable` directive within the `remote-control` statement block. For Knot DNS, configure a listen-on address and ACL or Unix socket path within the control statement block.

Configure server access control lists using iptables or its equivalent on your operating system to constrain update sources if you have a separate network interface on your server for management control than for DNS protocol transactions. Supplement these network interface controls with controls on DNS protocol transactions by transaction type to control access to DNS update functions on the otherwise broadly open DNS interface at the network interface ACL layer.

ISC BIND Controls   ISC BIND provides the following DNS transaction ACLs:

- allow-new-zones – configures whether zones can be added to the server at runtime using `rndc addzone` or deleted using `rndc delzone`. The default is no.
- allow-transfer – specifies an ACL on who can receive a zone transfer from this server. The default is any.
- allow-notify – accept Notify messages from hosts identified by the address match list in addition to corresponding zone masters. The default is to allow Notify messages from the configured zone master server(s) as configured in the masters statement of a given zone declaration.
- allow-update – specifies an ACL defining from whom dynamic updates will be accepted for slave zones which will in turn be forwarded to the zone's master server. The default is none which is appropriate for external authoritative DNS servers. For internal authoritative DNS servers, you may wish to enable updates from authorized DHCP servers.
- update-policy – enables specification of increased granularity of dynamic updates over and in lieu of allow-update to specify permissible (or denied) update sources, the scope of updates and even specific resource record types.

NLnet NSD Configuration   NSD does not support dynamic updates. This provides update security but it affords no mechanism for dynamic updates. For external authoritative DNS servers, you should not permit dynamic updates at all; perform all zone updates by configuring zone files on the master, then permit the notify/zone transfer process to propagate these updates to the slave servers. To control access to this process, NSD supports the following server daemon level access controls on DNS information update permissions.

- allow-notify – accept or block Notify messages from hosts identified with this directive by IP address or IP block with or without transaction keys.
- request-xfr – the listed [master] DNS server is queried for a zone transfer (IXFR/AXFR).

- provide-xfr – The listed address is allowed to request AXFR from this server. Zone data will be provided to the address. A transaction key may be specified for use during an absolute zone transfer (AXFR).

PowerDNS Authoritative Protection   The PowerDNS authoritative server supports ACLs for notifies, transfers, and dynamic updates. The following parameter settings for the Authoritative PowerDNS server apply.

- Dnsupdate = yes setting enables dynamic updates.
- allow-dnsupdate-from statement, for example `allow-dnsupdate-from 172.16.0.0/20, 192.168.128.0/17`. You can define from where updates may be accepted.
- updatepolicy enables you to define a Lua script to specify update authorization requirements.
- allow-notify-from limits from where a DNS NOTIFY message will be accepted. Also allow-unsigned-notify can be set to require TSIG-signed notifies only.
- allow-ixfr-ips enables specification of IP CIDR ranges from which AXFRs may be processed.

Knot DNS Sample Configuration   Knot DNS enables declaration of ACLs comprised of IP address blocks and the corresponding action enabled for the ACL: `update` in the case of dynamic updates, `transfer` for zone transfers, and `notify` for notifies. You may also define TSIG keys and require them with corresponding ACLs.

```
acl:
    - id: allow_update_acl
      address: 171.16.0.0/20
      action: update
```

As with BIND this ACL can then be applied globally or to specific zone(s).

```
zone:
    - domain: example.com
      file: example.com.zone
      acl: allow_update_acl
```

Defending Resolution Data in Motion with DNSSEC

Besides protecting your authoritative DNS servers' update processes in terms of dynamic updates, fully protecting DNS resolution data from falsification requires DNSSEC. From the authoritative DNS perspective, DNSSEC provides the best means of protecting the integrity of your namespace and is the definitive defense mechanism against cache poisoning attacks. As we introduced in the prior chapter, DNSSEC utilizes asymmetric cryptography and we reviewed the process for configuration of DNSSEC validation. In this chapter, we'll consider the DNSSEC zone signing process which assures the integrity of your namespace for validating resolvers. But first let's review the resolution process when DNSSEC validation is incorporated.

The DNSSEC Resolution Process   The basic resolution process we reviewed in Chapter 1 still applies in terms of locating the DNS servers that are authoritative for the domain name in question by walking down the domain tree, then querying an authoritative server for an answer. If the recursive server is configured to validate and the resolution data is signed, the recursive server will perform the validation once the resolution data has been received. Thus, the validation process commences after the resolution process completes though much of the data required for validation can be gathered during the resolution process.

Let's assume the recursive server is configured with the DNS root zone trust anchor. This trust anchor is the trusted key that ultimately validates signed resolution data down the chain of trust. Another way to think of this is given the resolution data received from the DNS server authoritative DNS server relevant to the query answer, we can trace the chain of trust up the DNS tree to the root zone. Figure 9.1 illustrates this basic process.

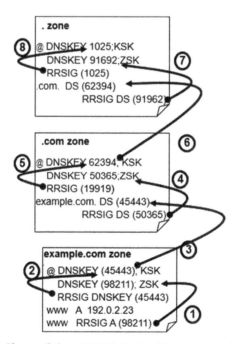

Figure 9.1.   DNSSEC Chain of Trust Traversal

Let's say my recursive server receives a query with a Qname of www. example.com, Qclass IN, and Qtype A. My recursive server is configured to validate DNSSEC and has the trust anchor corresponding to KSK id 1025 configured. My recursive server sets the DNSSEC OK (DO) bit in the DNS header and issues iterative queries down the domain tree. DNSSEC requires EDNS0 to support this extended Rcode field and also for the generally large response packets likely exceeding the 512 byte UDP packet limit. The example.com administrator has signed the zone and the authoritative DNS server responds with the answer to my query and includes a signature for the www.example.com A RRSet in the form of an RRSIG resource record.

To validate this signature, I'll need the zone's public key(s) that correspond to the private key(s) used to sign the zone. Zones are usually signed with two keys, a zone signing key (ZSK) to sign all RRSets in the zone and a KSK, a longer key which signs only the the zone's public keys. We'll discuss the motivation for this approach later but these public keys are published in the zone file in the form of the DNSKEY RRSet.

The recursive server first validates the query answer, the A record answer by hashing the answer and applying the ZSK to compare with the RRSIG signature. If this matches, shown as step 1 in Figure 9.1, the recursive server shall repeat this process to validate the ZSK and KSK by verifying the signatures on the DNSKEY RRset as step 2. Having performed these two validations, my recursive server can confirm that the query answer was signed by the example.com administrator and the answer was as published by the administrator. But do I trust example.com's key? No this key is not configured as a trust anchor.

Thus, the recursive server must determine if example.com is linked in the chain of trust to its parent zone, .com. This chain of trust linkage is published in DNS in the form of a Delegation Signer (DS) resource record. The DS record provides a hash of the corresponding child zone's KSK as a means to authenticate the child KSK. Step 3 in our process verifies the parent zone has a DS record corresponding to example.com's KSK. Step 4 validates the DS record's signature with .com's ZSK while step 5 validates .com's keys. Repeating these steps to the root zone in steps 6–8, we arrive at validating signatures for the root zone up to the KSK which has linked down to our query answer for www.example.com. Since we trust the KSK, we consider this answer as secure. While Figure 9.1 implies querying "up" the domain tree, typically the signatures, keys, and DS records are cached by the recursive server to expedite the validation process.

DNSSEC also provides authenticated denial of existence so if I had mistyped my query, the recursive server could authenticate the fact that the queried name does not exist in the zone as published by the zone administrator. This process relies on the NSEC or NSEC3 resource record types which provide a means of identifying the "gap" into which the query failed to prove answerable. If DNSSEC did not provide this feature, an attacker could attempt to poison resolver caches by sending NXDOMAIN responses for otherwise valid data. Imagine an NXDOMAIN answer honored

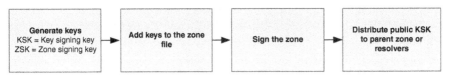

Figure 9.2.   Basic DNSSEC Implementation Steps

by caching servers for your www address. Fortunately, with DNSSEC this vulnerability can be controlled.

DNSSEC Zone Signing   Now that we've reviewed the validation process, let's examine the signing process. Signing a zone requires the generation of a key pair. As we discussed, the IETF's DNSSEC operational practices (58) recommends (but doesn't require) two key pairs: a ZSK which is a shorter key to lessen computational complexity and time for signing all of the zone resource record sets (RRSets). Typically a longer key, the KSK is used only to sign the DNSKEY RRSet, that is the RRSet that identifies the public keys, ZSKs, and KSKs, used (or to be or have been used) within the zone.

Figure 9.2 illustrates the basic process of signing a zone. Step 1 consists of generating two key pairs, a private and public pair as a ZSK and a KSK. For both pairs of keys for each zone, the public keys are published within the zone in the form of DNSKEY resource records as step 2. The third step utilizes the private keys to sign the RRSets in the zone. Again the KSK signs only the DNSKEY RRSet and the ZSK signs all RRSets.

The fourth step entails linking the KSK into the chain of trust by having the corresponding DS record provisioned in the parent zone. In the event that the KSK is a trust anchor for any resolvers, for example, if you've deployed an internal signed root or if your domain is not fully linked in a chain of trust to the Internet root, such resolvers must be updated to reflect the new trust anchor key.

Key Rollover   Once your zones are signed, keys must be rolled over occasionally. This typically creates some extra work for DNS administrators in having to perform this key rollover function. The fact that keys are rolled "occasionally" can lead to errors as is sometimes the case when performing complex tasks infrequently. Fortunately, most of these functions including key rollover can be automated thanks to leading vendor implementations as we'll discuss later in this chapter. Note that vendor implementations also enable you to manually manage zone signing and maintenance if you prefer that level of control. We will review the automated procedures here but offer references to further information for detailed manual implementation steps and troubleshooting tips.

The ZSK can be changed more frequently, for example, every 30–90 days, and such change has no impact on other DNS domains in the domain tree. The KSK however, is represented in the parent zone in the form of a Delegation Signer (DS) record which links the chain of trust up the domain tree. Hence the motivation for the KSK being a more secure (longer) key that may be changed less frequently, which is operationally preferable given the requisite update process in the parent zone to reflect the key change.

The process for key rollover mimics that of initial zone signing with the generation of a new key pair for the key being rolled, publication of the public key within respective zone file, re-signing of the zone, and for KSK rollovers, linking into the chain of trust. There is some added complexity however in that you cannot simply replace your current keys and re-sign the zone. Since validating resolvers cache not only resolution data but signatures and keys, we must account for the fact that a resolver may utilize a cached public key to validate signatures on a fresh resolution.

For example, if the recursive server queried for mail.example.com yesterday, which was signed, the recursive server would need to query for the zone's public key to validate the RRSet's signature. Let's assume the TTL on my public key (DNSKEY) record is 2 days. If I then query today for www.example.com, which also is signed, my recursive server must once again validate this RRSet's signature. Given it possesses a DNSKEY record for the zone in cache as the TTL has not yet expired, it shall attempt to use the cached key to perform validation. If the www RRSet had been signed today with a new key, the validation would fail and the resolution considered bogus. Recursive servers may possess the resolution data with signature or the public key or both within cache. The rollover process needs to account for a resolver possessing one or the other but not both to enable validation of the cached data based on what is currently published in the zone.

Thus, when rolling keys, you'll need to publish two keys for a given period of time, either signing with just one or signing with both. In the former case, the "prepublish" rollover method, two keys are included in the zone file, the incumbent key used to sign the zone and the new key which is published but not used in signing. Recursive servers seeking to validate signatures will obtain both keys within the DNSKEY RRSet will and try both during the validation process; in this case, the incumbent key validates. If the new key had also been used to sign the zone, each RRSet in the zone would have two signatures, corresponding to the two keys. This double signature approach would vastly increase the size of your zone files and resolution data payload given two signatures per RRSet. Thus, the prepublish method is typically recommended to roll ZSKs so that each RRSet has one signature. And because the KSK signs only the DNSKEY RRSet, and because the parent zone's corresponding DS rollover must coincide, the dual-signature approach is typically used to rollover KSKs.

PREPUBLISH ROLLOVER   Let's dig deeper into these two rollover strategies by first considering the prepublish rollover method of Figure 9.3. Our initial condition features our zone signed with a ZSK with key id 14522 and KSK key id 6082. Both of these keys are used to sign the zone as indicated by the pen icon adjacent to the

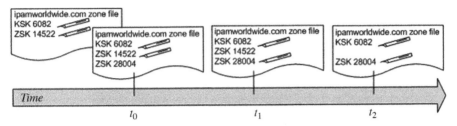

Figure 9.3. Prepublish Rollover

corresponding keys in Figure 9.3. Let's initiate the rollover process at time t_0, by publishing a second ZSK DNSKEY resource record with key id 28004 into the zone file. Our DNSKEY RRSet now consists of these three keys. With this change, we re-sign the zone with KSK 6082 and ZSK 14522.

The new ZSK (28004) is not currently signing the zone but the key is made available for resolver and recursive server caching. As such we need to wait an interval of time approximately equal to the amount of time required to distribute zone updates from the master server to all of its slaves (upper bounded by the zone expiration time) plus the TTL of the DNSKEY RRSet. When this time elapses, we reach time t_1, and now we can re-sign the zone, keeping all three keys in the DNSKEY RRSet, but now signing with KSK 6082 and ZSK 28004.

We need to retain the old ZSK in the zone for a time so that resolvers possessing resolution data with signatures from the old key, for example, fetched right before time t_1, can still be considered valid. Thus, we should keep the formerly signing ZSK in the zone file for an interval of the time required to distribute zone updates from the master to the slaves plus the maximum TTL value of zone data. When this time elapses, we reach time t_2, and we can remove ZSK 14522 and re-sign the zone.

In some instances, it may be simpler from an operations perspective to always publish two ZSKs, one active and one either being staged or pending departure. Thus, at time t_2, we could introduce a third ZSK which would eventually be used to sign the zone upon the next rollover. This third ZSK would remain published until the next rollover time (e.g., 30–90 days usually) which maps to time t_1 when the new ZSK will be used to sign the zone. The $t_2 - t_1$ interval should retain the same time period as above and at t_2, the old ZSK can be replaced by a new ZSK to be used to sign the zone upon the next rollover.

DUAL-SIGNATURE ROLLOVER   The dual-signature method is typically recommended for KSK rollovers and the basic process is illustrated in Figure 9.4. Because the parent zone's DS record must reference a valid KSK in this zone to link the chain of trust, we'll illustrate the state of the parent zone DS record in Figure 9.4. Our initial condition is as before with a KSK with key id 6082 and ZSK with key id 14522. The parent zone DS properly references the active KSK 6082.

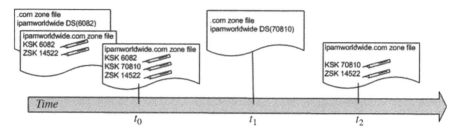

Figure 9.4. Dual-Signature KSK Rollover

We begin the rollover process by creating a new KSK key pair and publishing the public key in the zone file. Sign the zone with both KSKs (and the ZSK, of course). Next, we need to inform our parent domain administrator that we're rolling our ZSK and the corresponding DS record in the parent zone must be updated to reflect this. The method(s) of performing this update will be dicated by your parent zone administrator's policies. You may need to securely login to a web portal and upload your DNSKEY or the corresponding DS RRSet. Or you may be able to upload it directly. Another mechanism defined by the IETF in RFC 7344 (59) calls for publication of the change in your DNS zone though the publication of a CDS and/or CDNSKEY resource record in your zone. The parent zone may periodically poll its children zones for the existence of one or both of these records as a signal to update its corresponding DS record(s). An out of band notification mechanism defined by the parent zone administrator may also be used to initiate the DS record update.

Once you've confirmed the parent zone has published the DS record corresponding to your new KSK and the longest zone TTL has expired since signing with both KSKs, the old KSK (14522 in our example) may be removed from the zone and the zone re-signed. Note that if your KSK is configured as a trust anchor within recursive resolvers which utilize RFC 5011 (50) for automated trust anchor management, you'll need to set the "revoke" bit on the outgoing KSK to signal its outgoing state to such resolvers. Figure 9.4 would be modified in this case at time t_2 where both KSKs would remain published and signing the zone, but the KSK 6082 would be published with its Revoke bit set for a period of time equal to the maximum zone TTL, after which KSK 6082 may be removed.

Algorithm Rollover   As cryptography technology evolves and new signature algorithms are introduced into DNSSEC standards, you may desire to add and/or remove algorithms used for signing your zones. Note that validating resolvers will also need to support one or more algorithms that you've implemented to enable proper validation.

The algorithm rollover process entails prepublication of the signatures of a new set of keys using the new algorithm prior to publication of the keys themselves. The reason for this stems from the requirement that every RRSet have a valid signature

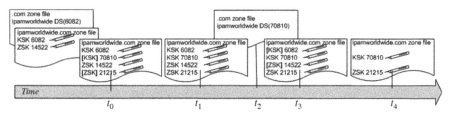

Figure 9.5.   Algorithm Rollover

for every algorithm represented in the zone's DNSKEYs. Considering Figure 9.5, we start with KSK 6082 and ZSK 14522, with our parent DS properly referencing our KSK.

At time t_0, we begin by signing the zone with the private keys utilizing the new algorithm and publishing the corresponding signatures but not the public keys. We denote this in Figure 9.5 as enclosing the KSK or ZSK text within square brackets with the pen indicating publication of respective signatures. Thus the private keys sign the zone data but the public keys are not published as yet. This enables resolver caches to obtain signatures with the current and new keys. If the keys had also been published at this time, resolvers could fetch the DNSKEYs and previously cached resolution data would not have signatures yet for the new keys, violating the requirement.

Once the zone master–slave propagation time plus zone TTL time has transpired, the corresponding DNSKEY records may be added to the zone at time t_1. By now new resolution data with both sets of signatures should be cached. At time t_2, after the DNSKEY TTL expires, the parent zone administrator may replace the DS record to reference the new KSK 70810. Once this DS(70810) is published in the parent zone, the keys using the former algorithm may be removed once the DS TTL expires at time t_3. Note that the signatures for the removed keys should remain in the zone until once again the zone propagation plus TTL time is reached. At this time, t_4, the signatures generated with the old keys may be removed. If your KSK is a trust anchor, insert step $t_{2.5}$ after the DS record is update to revoke the outgoing KSK for a DNSKEY TTL time period before removing the KSK.

You can also use this algorithm rollover process to rollover from NSEC to NSEC3 authenticated denial of existence. These records authenticate the non-existence of queries to a signed zone with NSEC publishing the "next secure" record in the zone, while NSEC3 publishes hashed next records to hinder simple zone footprinting.

Key Security   The private keys corresponding to your published public keys must be secured from theft. Should an attacker obtain a private key currently in use for signing one of your zones, he or she could sign arbitrary DNS resource records and sign them with your private key, successfully poisoning validating resolver caches. Hardware security modules (HSMs) may be deployed to securely store private keys and to perform zone signing using the PKCS #11 cryptographic token interface (60).

Emergency rollover procedures should be devised and documented in the event of compromise of a private key corresponding to an active KSK or ZSK. Should an attacker obtain the private key, he/she could forge zone data and sign it with the private key. Resolvers and recursive servers would authenticate the falsified data based on the published corresponding public key. As we've seen, the ZSK can be changed autonomously and should be changed immediately. This may cause validation issues due to conflicting cached data. Note that your zone is still vulnerable until the TTL of the signatures generated by the compromised key(s) expire.

Changing the KSK, however, does require broader involvement and coordination. We recommend documenting a process for emergency rollovers that includes the parent zone administrator, as well as a means to communicate to users who have configured the KSK as a trust anchor. This could be via a registered email list or secure website posting. There are three ways you can perform an emergency KSK rollover, each with its corresponding risks.

1. Maintain the chain of trust

    a. Generate a new KSK and add the corresonding DNSKEY record to the DNSKEY RRSet, keeping the compromised key published. Lower the DNSKEYs' TTL value to promote rapid expiration during this rollover.

    b. Sign the DNSKEY with both the new and compromised KSKs. The reason we continue publishing and signing with the compromised KSK is due to the fact that the parent zone still references the compromised KSK via a DS record. Set the signature validity interval to the time until the parent zone can publish the DS corresponding to the new KSK plus the DS TTL value.

    c. Upload the DS record corresponding to the new KSK to the parent zone administrator and request removal of the DS record pointing to the compromised KSK.

    d. After the new DS record appears in the parent zone and the old DS expires from caches based on its TTL, remove the compromised KSK from the DNSKEY RRSet and re-sign. Note that you may have to repeat step b to refresh signatures if this DS publication process is delayed.

2. Break the chain of trust

    a. Publish a new KSK and remove the old KSK from the DNSKEY RRSet and re-sign the zone. Note that the parent zone still points to the compromised KSK and not your new KSK, enabling the attacker and not you to validate up the chain of trust.

    b. Upload the DS record corresponding to the new KSK to the parent zone administrator and request removal of the DS record pointing to the compromised KSK.

    c. Once the DS for the new key is published, the chain of trust will have been repaired though the compromised key will likewise still validate while the old DS remains cached.

3. Render the zone insecure

    a. Request removal of the (all) DS records from the parent zone. This will break the chain of trust and render your zone insecure. This will also render the attacker's zone insecure with its use of the compromised KSK.

    b. After the DS TTL expires, generate a new KSK and add the corresponding DNSKEY record to the RRSet and sign the zone.

    c. Upload the DS record corresponding to a new KSK to the parent zone administrator.

Vendor Implementations   Most leading vendors provide some level of zone signing automation features that can simplify your implementation and ongoing management of DNSSEC. Nevertheless, you must monitor your implementation and maintain emergency rollover procedures in the event of a private key compromise.

INTERNET SYSTEMS CONSORTIUM   As of BIND 9.11, ISC BIND fully automates the DNSSEC signing and maintenance functions with a new `dnssec-keymgr` Python wrapper (61). You can define your DNSSEC policy in a configuration file, specifying key parameters, key publication and rollover policies, and apply for all zones or set different policies for different zones. Please consult (7) and (51) for further details on BIND configuration for DNSSEC.

NLNET NSD   NSD supports secure zone resolution but does not automate DNSSEC key generation or zone signing. NSD can load and resolve pre-signed zones.

POWERDNS   The PowerDNS authoritative server supports in-line signing of DNS resolution responses and pre-signed zone file like resolution. The `pdnsutil` utility (62) enables the generation, activation, deactivation, disabling, export and import of keys, and various zone management commands. This utility enables you to create, stage, activate, and roll keys. You can simply sign a zone using the `pdnsutil` `secure-zone` command specifying the zone to be signed.

KNOT DNS   Knot DNS provides manual or automated zone signing based on a signing policy defined in the configuration file (63). The `keymgr` utility can be used to generate and activate keys manually. In automated mode, new keys are generated based on the signing policy for the zone and ZSKs (not KSKs) are automatically rolled. A signing policy and zone association might look like the following:

```
policy:
  - id: secure-policy
    algorithm: RSASHA256
    ksk-size: 2048
    zsk-size: 1024
```

```
zone:
  - domain: example.com
    dnssec-signing: on
    dnssec-policy: secure-policy
```

SUMMARY

This chapter described securing DNS authoritative data by securing the update process for authoritative data. Such security is required for both in-house DNS servers as well as external DNS service providers. We also described signing your zone data using DNSSEC. DNSSEC secures your resolution data to enable resolvers to validate it as published by you. You should document your signing plan to define your DNSSEC policies including:

- Key lengths for ZSKs and KSKs
- Enumerate zones to be signed
- Signature intervals
- Key rollover intervals

Also, define procedures for the following activities.

- Key rollover procedures
- DS publication procedures to interwork with each parent zone(s)
- Emergency key rollover procedures

10

ATTACKER EXPLOITATION OF DNS

INTRODUCTION

As we've seen, the DNS is fundamental to the proper operation of nearly every IP network application, from web browsing, email, to multimedia applications, and more. Its essential function and decentralized architecture serve to attract attackers seeking to exploit its decentralized architecture and rich data store for sinister activities. Attackers may also leverage and use DNS given its ubiquity and its general permissibility to flow freely through networks, exposing networks to attacks that leverage this freedom of communications.

Major vulnerabilities exposed by the necessary free flowing of DNS traffic fall into two main categories: network reconnaissance and tunneling. A third category, the use of DNS by malware for "normal" resolution will be covered in Chapter 11.

Network Reconnaissance

DNS by design contains a repository of hostname to IP address mapping among other things. Many attacks in general begin with a reconnaissance phase to enable an attacker to identify a worthy target. "Worth" in the attacker's mind may vary depending on the attacker's objectives. Bored college kids could attempt to bring down

DNS Security Management, First Edition. Michael Dooley and Timothy Rooney.
© 2017 by The Institute of Electrical and Electronic Engineers, Inc. Published 2017 by John Wiley & Sons, Inc.

a website merely for bragging rights among friends, to change the outcome of an election, or to otherwise achieve "fame." Attackers seeking financial gain may desire to target systems appearing to hold personal or financial data. State sponsored attackers may seek targets containing intelligence or classified materials. If an attacker desires to glean information about particular hosts that may be more attractive to attack than others, he/she may start with DNS.

- Name guessing – One brute force approach to such reconnaissance consists of guessing hostnames of interest (e.g., "payroll") and issuing standard DNS queries to obtain corresponding IP addresses if they exist.
- Zone transfers – Impersonating a DNS slave server by attempting to perform a zone transfer from a master is a form of attack that attempts to map or footprint the zone. That is, by identifying host to IP address mappings, as well as other resource record information, the attacker attempts to identify targets for direct attacks.
- Zone footprinting – Leveraging DNSSEC records to hop through a signed zone to identify provisioned resource records. The "next secure" NSEC and NSEC3 resource records identify the next resource record set within a signed zone in order to authentically deny existence of names in between.

Data Exfiltration

Data exfiltration refers to the transmission of data originating from within one security domain, for example, an enterprise network to another entity or organization. There are two basic forms of data exfiltration using DNS.

- DNS as data protocol (tunneling) – DNS tunneling entails the use of the DNS protocol as a data communications protocol. This technique enables a user or device within the network to communicate with an external destination, easily traversing firewalls (DNS is generally permitted through firewalls).
- DNS as resource locator – an attacker may attempt to install malware on devices via phishing or other attacks that bait users into opening executable email attachments or installing software from an attacker website. Whether a device is attacked while inside the enterprise network or an infected user device is physically brought onto the network, if they are trusted within the confines of an enterprise network, they may have access to sensitive information. The malware may perform data collection, locating internal resources using DNS reconnaissance. In addition, DNS could be used to identify the current IP address of the attacker's external destination for exfiltration of the information.

DETECTING NEFARIOUS USE OF DNS

When attackers leverage DNS to query for network information, its intended purpose, how can one detect such activities? Clearly one cannot detect the intent of someone issuing queries but certain query patterns may raise suspicions of intent.

Detecting Network Reconnaissance

If one of your DNS servers receives a query requesting a zone transfer from an IP address that is not that of one of your DNS servers which are authoritative for the same zones, a potential attacker could be seeking naming and IP address information for your network. You should define ACLs on zone transfer transactions such that any such attempts for information can be prevented and logged. Any detected attempts for zone transfers should be investigated to attempt to identify the requesting source. You can also specify shared secret transaction keys which provide additional authentication of zone transfer requests and inhibit such requests that use spoofed IP addresses.

An attacker may attempt to issue an ANY query to retrieve all resource records for a given domain name within your zone. This may be helpful in identifying certain applications running on servers of a given domain name. While you may not desire to prohibit ANY queries, you should audit your query logs to identify the source of ANY queries to determine if any suspicious traffic subsequently originated from the query source's IP address. Some implementations like BIND enable you to configure your DNS server to respond to ANY queries with a meager subset of zone information using the `minimal-any` option.

Failing attempts to zone transfer zone information, an attacker may attempt to obtain zone information by footprinting the zone if it is DNSSEC signed. The NSEC/NSEC3 resource record types identify the next canonically ordered resource record set within a zone, enabling one to successively query through the zone file to identify its resource records. NSEC records publish the next RRset owner name plainly, while NSEC3 publishes a hashed version, encumbering but not inhibiting footprinting attempts. A new NSEC5 record has been proposed within the IETF which uses asymmetric key pair hashing to more effectively prevent footprinting.

Footprinting zones leveraging DNSSEC records are more efficient than pure name guessing, but either method may indulge attackers with prospective hosts to target. Given that your DNS information is freely exposed to those within the networks permitted to query, which may include attackers, review query logs periodically to identify a high rate of queries from a given IP address or processed ANY queries. These logs may indicate a footprinting attempt but by no means proves it. If your intranet site, for example, contains numerous images or links that browsers organically resolve, such a flurry of queries would be expected. A surge of queries, perhaps an order of magnitude higher than such a normal flurry should trigger an alert at least for investigation into the presence of a footprinting attempt or perhaps an inefficiently coded webpage. Periodically review network activity logs to determine if any suspicious traffic subsequently originated from the query source's IP address.

DNS Tunneling Detection

Data exfiltration may be accomplished by either using DNS as the data exfiltration transport protocol or by using DNS to locate an Internet destination to which to exfiltrate data or both. We will consider the former case in this chapter, while we'll cover the latter in Chapter 11.

DNS tunneling entails transmission of data payload via the DNS protocol. The tunnel endpoints comprise a DNS client behaving as a resolver presumably within the organization with data to be exfiltrated, and a DNS server which terminates the tunnel and ultimately forwards the data to the intended destination. The DNS server, authoritative for a given domain name must be provisioned within the domain tree, for example, with parent zone NS/glue records. The recursive server to which the stub resolvers issue the query locates the authoritative server then issues the "query." Data is encoded within the DNS query and the server can acknowledge and send additional information via its responses. A tunneling session must be originated from the client when using a standard resolver, for example, that residing on the client machine, as the resolver does not listen for unsolicited query answers.

The client encodes the data to be transmitted using base32, base64, binary, hex, or NetBIOS encoding as the "hostname" label, suffixed with the tunnel server's authoritative domain name. Thus, the Qname is formed by encoding application data, most commonly a web browser http request, with base32 or base64 to restrict resulting label characters within the bounds permitted by the DNS protocol. Let's say, for example, an attacker desires to copy a file using HTTP. The HTTP PUT transaction might start with this portion of the HTTP header.

```
PUT doc/stolen/examplecorp HTTP 1.1
Host: www.tunnel-example.com
Accept-Encoding: gzip
Content-Length: 27401
```

which the tunnel software encodes into an A record query with Qname, Qclass, and Qtype.

```
KBKVIIDEN5RS643UN5WGK3RPMV4GC3LQNRSWG33SOAQEQVCUKAQDCL
RRBJEG64.3UHIQHO53XFZ2HK3TOMVWC2ZLYMFWXA3DFFZRW63IKIFR
WGZLQOQWUK3TDN5SG.S3THHIQGO6TJOAFEG33OORSW45BNJRSW4Z3U
NA5CAMRXGQYDC===.tunnel-example.com.  IN A
```

This query name may be included in an A, AAAA, TXT, or other query type and issued to the local recursive server. Upon locating the authoritative server and issuing the query, the response would contain the encoded response data within an answer with the queried type or CNAME for an A/AAAA query.

Several DNS tunneling utilities have been developed and are freely available on the Internet, including the following (64):

- dns2tcp – Linux server and Windows client; supports KEY and TXT RRTypes
- DNScat-P – Java-based software runs on Unix systems; supports A and CNAME RRTypes
- DNScat-B – encodes requests in either NetBIOS or hex encoding and uses A, AAAA, CNAME, NS, TXT, and MX RRTypes
- DNScapy – supports SSH tunneling over DNS including a Socks proxy; configurable to support CNAME, TXT, or both RRTypes
- Heyoka – encodes requests in binary and leverages EDNS for larger payloads; also uses source IP address spoofing to spread transactions over seemingly diverse tunnel endpoints to avoid detection
- iodine – multiplatform program runs on Linux, MAC OX S, Windows and has been ported to Android
- OzymanDNS – sets up an SSH tunnel over DNS for file transfer; requests are base32 encoded and responses are base64 encoded TXT records
- psudp – uses a broker server through which tunneled packets are sent for holding until a corresponding request is received
- NSTX – "name server transfer protocol" runs on Linux
- tcp-over-dns – Java-based server and client that runs on Windows, Linux, and Solaris; supports LZMA compression and both TCP and UDP tunneling
- VPN over DNS
- squeeza – an SQL injection tool that supports exfiltration via http errors, timing, and DNS; for DNS tunnels, data is encoded in the FQDN used in the request (Qname)
- TUNS – a Ruby application that uses only CNAME records
- DeNiSe – Python application for tunneling TCP over DNS

Methods for detection of DNS tunneling include the examination of DNS payload and of DNS transactions. These methods can also be applied to detecting queries to random domains using dynamically generated algorithms (DGAs) commonly used for malware detection evasion as we'll discuss in Chapter 11. The SANS Institute published a paper regarding the detection of DNS tunneling using Splunk (65).

DNS Payload Analysis   DNS payload analysis entails assessment of the message size for DNS requests and responses. Requests or queries include a question section with a resource record-formatted query with Qname, Qclass, and Qtype.

In a DNS tunneling scenario, the client will likely attempt to pack in the maximum amount of data per packet to offer reasonable response time to the user or application. Qnames with labels at or near the maximum of 64 bytes and total Qname length near or at 255 bytes may be suspect.

The entropy of the Qname hostname label may likewise raise suspicions about DNS tunneling. Most hostnames are meaningful and dictionary based unlike encoded data, though this is not the case with hostnames not intended for human entry like images or content delivery network links. Other forms of hostname signature analysis such as the ratio of numerical digits versus letters, or alignment to known tunneling signature patterns can be used to detect tunneling.

The use of uncommon Qtypes could also indicate tunneling. Appendix B lists the set of currently defined types though most are not seen in practice in most networks. If you track queries over time, you may be able to detect uncommon Qtypes for your network. Use of a Qclass other than IN could also provide a tip-off to the existence of a DNS tunnel.

Certain tunneling software may exhibit deterministic characteristics (like a signature) that could be identified if DNS packets are monitored and analyzed. For example, possible NSTX tunneling over DNS may be indicated by the presence of the characters, "cT" within the first three characters of the Qname.

DNS Traffic Analysis   A tunnel implies two fixed endpoints, at least for the duration of a tunneling session, between which DNS packets are exchanged. Thus, a tunnel could be indicated by a relatively high volume burst of traffic between a particular client and an external DNS server, which is authoritative for the domain suffix of each query. If correlation of bidirectional DNS transactions is difficult, you could monitor high DNS traffic from a given IP address (client) or for a given domain within the Qname. A large number of unique hostnames being queried for a given domain may also trigger suspicion. Encoded data will likely produce unique Qname labels prepended to the tunnel endpoint domain suffix. An unusually large burst of NXDOMAINs could be indicative of DNS tunneling, particularly the Heyoka implementation (see below), though this metric could also indicate other attack forms such as the bogus queries DoS attack.

Some tunneling utilities may attempt to bypass the local recursive server and perform recursion directly, behaving as a recursive server. The goal here is to avoid detection should the recursive servers be monitored. However proper recursive server deployment with associated firewall rules constraining the source IP addresses of outbound DNS requests to deployed caching servers can mitigate such attempts.

The geographic location of the authoritative DNS server could indicate tunneling if it resides in a region where few prior queries have been issued historically. Also, if the age of the domain name suffix for which the DNS server is authoritative is relatively small, for example, this is a new domain, the domain could be considered suspicious or at least cautionary. Oftentimes attackers will procure a new domain for the express purpose of conducting an attack such as for provision of a DNS tunnel endpoint or as a malware command and control center.

If you are able to correlate DNS requests with application activity, the presence of DNS activity without subsequent application activity may represent use of DNS for tunneling. For example, a DNS query will typically be issued by a browser to locate a web resource; if an HTTP request is absent following such a request, perhaps DNS was being used for nonresolution purposes.

MITIGATION OF ILLICIT DNS USE

Mitigating attacker use of DNS to resolve hostnames is challenging because DNS is merely performing its intended function and usually analysis of multiple DNS transactions over time is required for detection.

Network Reconnaissance Mitigation

The following tactics may be configured to reduce the ability of attackers to reconnoiter your zone information.

- Define zone transfer ACLs as a key defensive tactic against illicit zone transfer requests.
- Log zone transfer requests and review logs periodically to identify any requests from IP addresses outside of the set of authoritative DNS servers.
- Require signatures on zone transfers for added authentication of DNS servers requesting zone transfers.
- While you can define ACLs on what IP addresses can query your servers, such a control is not applicable to external DNS servers. For external servers, publish only your externally reachable hostname and IP addresses, such as your web and email servers.
- Never publish internal hostnames on your external servers.
- For partner links, deploy independent DNS servers to publish only partner-reachable host information.
- You should define ACLs on internal DNS servers to constrain the address space from which queries will be processed.
- You could also go so far as to deploy separate servers or use DNS views to restrict resolution of certain highly sensitive hostnames.

Mitigation of DNS Tunneling

The following tactics may be implemented to facilitate DNS tunneling detection. Upon detection, investigation into the tunnel endpoints can lead to disabling the tunnel.

- Scanning your host systems within your network for the presence of known tunneling utilities periodically can help proactively identify prospective tunnel endpoints. Other DNS tunneling tools may emerge or be encompassed within malware.

- Analysis and correlation of DNS traffic to identify a relatively bursty exchange of queries and answers among a pair of IP addresses, client and server, each with large QNAME and RDATA payloads, respectively, can prove useful in detecting a DNS tunnel. Detection of tunneling using a combination of the techniques we discussed in the DNS Tunneling Detection section can help detect possible tunnels in operation.
- Isolation of the host device as the tunnel endpoint can allow you to mitigate the threat through investigation and removal of the suspect code.
- More closely monitoring the tunnel domain DNS server can provide forensic data for use in characterizing the attack for prevention in future.
- Blocking the domain using a DNS firewall approach can effectively tear down the tunnel in question.

11

MALWARE AND APTS

INTRODUCTION

Malware has grown to become a menacing force in enterprise networks. In earlier days, malware consisted of malicious software that stealthily installed itself on a device to perform a pre-programmed form of attack. Unfortunately, this static form of malware is a rarity today, and malware is growing increasingly more sophisticated so as to hide itself on host systems, operate stealthily to avoid detection and remediation, and contact external command and control (C&C) centers for new software and instructions. Such malware effectively transforms the host machine into a bot for remote use by the attacker, and such malware installed on multiple devices can be formed into a botnet with which the attacker may launch a variety of attacks from multiple endpoints.

Stealthy resilient malware is considered an advanced persistent threat (APT). The malware is advanced in its ability to adapt with software updates from the C&C center, persistent in the sense that it utilizes a variety of strategies to avoid detection and thus persist on the network, and a threat given the attack forms range from DDoS to data exfiltration. APTs often utilize DNS to locate the C&C center. After all, if the malware used a hard-coded IP address, the malware could be shut down by simply blocking the corresponding IP address once the malware has been detected.

Using DNS enables the malware operator to modify their IP address to avoid detection. They often modify their domain name as well to avoid detection,

DNS Security Management, First Edition. Michael Dooley and Timothy Rooney.

creating various forms of fast-flux networks using dynamically generated domain names. We'll discuss this later in this chapter.

MALWARE PROLIFERATION TECHNIQUES

With the ubiquity of mobile devices which employees bring into the office and connect to your network, your control of endpoint security may be very constrained. Your perimeter defenses may be effective for protecting your network devices from attacks originating from the Internet, but if users bring infected devices physically into the confines of your network, you are susceptible to internal attacks instigated by malware unwittingly installed on user devices.

But malware can also be installed on residential or nonuser devices which may be less protected from Internet-based attacks. The September 2016 DDoS attack on Brian Krebs' security blog leveraging the Mirai malware installed on Internet of Things (IoT) devices highlights this vulnerability (7). A similar attack was launched in October 2016 against DynDNS (66) weeks later where over 100,000 IoT devices infected with the Mirai malware launched a DDoS attack exceeding 1.2 Tbps. The botnet was formed by leveraging default user names and passwords installed on IoT devices such as surveillance cameras and home routers to gain access and install the malware.

Besides leveraging manufacturer default user IDs and passwords to access and install malware on devices, other methods for installing malware are widespread including the following tactics.

Phishing

Attackers may send generic emails enticing users to click a link that leads to their website for installation of their malware. Such emails may promise fortunes for the claiming or inform readers of a need to validate personal information or otherwise react to a pending dangerous situation, or any number of varieties of enticement. The objective is to formulate an email that appears credible enough that several users will click your embedded link in order to download and quietly install their malware.

Spear Phishing

Spear phishing is a more focused attack where the attacker targets a specific individual with the intent to appear "familiar" when contacting the target individual. By researching social media posts, public Internet information or social engineering, the attacker may be able to send an email that garners a click from the target to download and install the attacker's malware.

Downloads

Who wouldn't want free software? Such offers of free stuff certainly attract attention and can be a useful tactic in phishing related attacks. But any time users download software (even free malware removal software), music, games, etc. from Internet sites, they are susceptible to malware installation. Virus (malware) protection software can help to identify and prevent installation, but new malware may not yet be recognizable by virus protection as we shall discuss later.

File Sharing

Any form of file sharing using shared media, network drives, or network protocols like FTP, HTTP, or SCP exposes the recipient system to malware installation. At minimum, virus protection scanning of incoming files or of the drive can help prevent installation of known malware. Users must be trained to utilize installed virus protection, or better yet, the protection should run automatically without user intervention.

Email Attachments

Clicking innocent looking email attachments may utilize embedded macros to install malware on the corresponding machine. Locky ransomware typically arrives at a target machine via spam email in the form of a Microsoft Word (or Excel) attachment. When the attachment is opened and macros are permitted, the macro runs to download the malware and lock the files on the victim machine while displaying a ransom message.

A variant on this form of attack is subtler with the attacker embedding OLE (object linking and embedding) objects within a Microsoft document. When a user clicks on the embedded Word document, for example, the document opens and runs a macro or embedded code to install malware.

Watering Hole Attack

A watering hole attack targets visitors of a given website. The website may function as a community of interest destination for users the attacker seeks to target. An attacker who can successfully infect the website to install his/her malware and entice visitors to the website to download it can successfully infect devices of such visitors.

Replication

Once installed on a device, some malware snoops your network and replicates itself, installing on other devices. The malware can snoop IP traffic on the wire to identify potential targets to which to attempt to replicate.

Implantation

The Mirai botnet, which attacked krebsonsecurity.com and DynDNS as mentioned previously, was formed when the attacker infiltrated a large number IoT devices such as surveillance cameras and other Internet-addressable unmanned "things." For the most part, the attacker hacked into these IoT devices by merely logging in using vendor default user IDs and passwords. Having gained access, the malware was installed, establishing a huge botnet from which attacks can be launched. The attacker then published the source code of this Mirai malware online for free use by like-minded attackers to leverage this strategy.

Malware Examples

The following is merely a sampling of malware that has been identified and characterized. The complete list is much larger and continues to grow.

Backoff (67) – malware targeting point-of-sale (POS) devices to capture credit card information to exfiltrate payment information to its C&C center. Backoff malware had infiltrated some high profile retail companies.

Cryptolocker (68) – ransomware generally targeted at Windows machines via an email attachment typically appearing as a zip file containing a file with a .pdf extension which executes when opened.

Dridex (69) – P2P malware which seeks to infect computers, harvest credentials, and steal money from users' financial accounts, as well as participate in DDoS attacks and send spam.

Locky (70) – ransomware which is typically installed by virtue of a user opening an email attachment which triggers the running of document macros.

Masque (71) – malware installable on iOS devices that can steal login credentials, access sensitive data from local cache, and gain iOS root privileges.

Mirai (72) – publicly published malware that installed itself on tens of thousands of IoT devices forming a botnet, launching high profile attacks on Krebs Security and DynDNS.

Nivdort (73) – malware that can delete Microsoft Windows system files, change security settings, and corrupt the Windows registry.

Simda (74) – botnet with self-propagating capabilities which may reroute a user's Internet traffic to attacker websites, obtaining user credentials, installing additional malware, or performing other malicious activities.

Zeus or GameOverZeuS (68) – Trojan horse malware that runs on Microsoft Windows targets often used to gather keystrokes or form captures to steal financial information.

MALWARE USE OF DNS

DNS Fluxing

Use of DNS enables the attacker to change IP addresses and domain names quickly to avoid detection. To avoid rapidly changing the IP address of the C&C center itself, the botnet administrator may enlist a set of bots to serve as proxies. Proxying provides a layer of abstraction to hide the C&C center IP address. The attacker rapidly changes IP address associations in order to prevent reactive blocking of these IP addresses by security personnel. If such an IP address is implicated, the attacker likely has already moved on to a new set of IP addresses as proxies for the C&C system.

The means of moving these IP addresses consists of the attacker configuring a set of bots to serve as proxies to the C&C center, let's say reachable at the domain name cc.example.com. The attacker has registered the example.com domain and operates DNS servers to resolve its namespace, including cc.example.com.

Consider Figure 11.1. On the left side of the figure, at time t_1, the botnet administrator has selected a pair of proxy bots and configured their respective IP addresses in the example.com zone file. The bot on the left queries the example.com DNS server to receive resolution data pointing to 192.0.2.62 or 192.88.99.5. The bot attempts to connect to one of these IP addresses, 192.88.99.5 in this example, and connects to this bot, which proxies the connection to the C&C center. The resolution data (resource records) to the cc.example.com host are configured with a very short TTL within the resource record, of the order of 5 minutes or less.

If after expiration of the TTL the bot needs to connect again, it (its recursive server) will no longer possess the cached response and will once again query the example.com DNS server, say at time t_2, at the right of Figure 11.1. By now the botnet administrator has modified the resolution data to now direct queries to cc.example.com to 192.0.0.207 or 198.18.0.193. The malware connects to the C&C center via one of these proxies.

This rapid changing of the C&C domain's A or AAAA records for the botnet controller system is referred to as *fast fluxing*. With bots and the DNS servers under control of the botnet administrator, detection of the C&C IP address is challenging; though if you can identify the cc.example.com DNS server address as the attacker's DNS server, you could shut down communications by blocking queries to that DNS server, though this may render shutting down of valid domain queries if the attacker is using a public DNS service whereby numerous domains are hosted on a given set of DNS servers.

To improve stealth, some attacks also configure certain bots to perform the functions of a DNS server, providing resolution for the botnet's C&C proxies. Figure 11.2 illustrates this scenario. When a bot issues a query for cc.example.com, the respective recursive server may start at the root DNS server, refer down to the TLD (.com), and then to the botnet administrator's domain, example.com. Fluxing of authoritative DNS servers for the example.com domain requires the attacker to rapidly modify its NS and glue resource records for the example.com domain. At time t_1, on the left of Figure 11.2, for example, the NS records (actually the glue records) refer to a pair of authoritative name servers for example.com, namely 223.255.255.7 and 191.255.0.55.

The recursive server then queries one of these "DNS servers," in this case themselves bots, which then resolve the cc.example.com query to its IP addresses, 192.0.2.62 or 192.88.99.5. The bot then connects to the C&C center via one of these

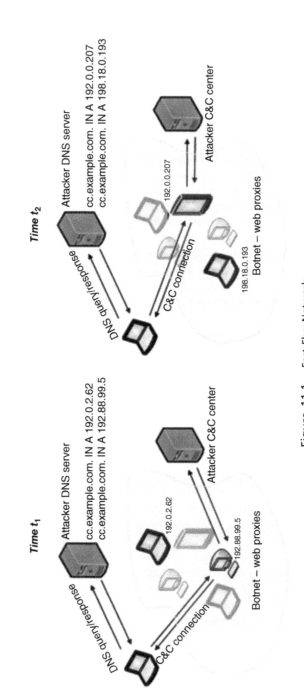

Figure 11.1. Fast-Flux Network

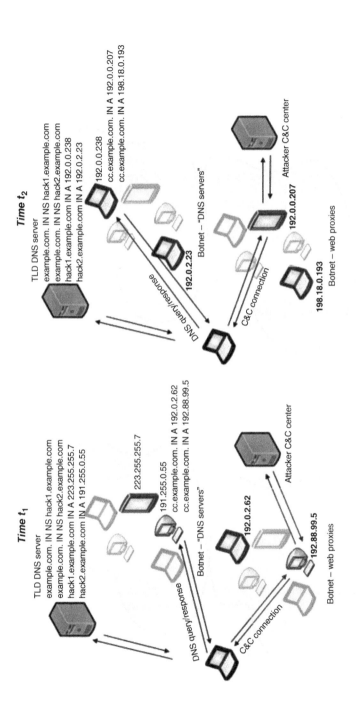

Figure 11.2. Double-Flux Network

proxies. The glue records in the TLD domain utilize a short TTL as well to time out the resolution of the botnet's DNS server IP addresses.

By time t_2, the bot's recursive server's cache for the cc.example.com destination as well as the example.com NS and glue records have expired, prompting a query to the TLD (.com) servers. The glue records now reflect different IP addresses, 192.0.0.238 and 192.0.2.23. When queried for cc.example.com, either will respond with a new set of botnet proxy IP addresses for cc.example.com. Given the dynamic IP addresses of not only the C&C center but its DNS servers, this form of fast flux is referred to as *double flux*.

Fast-flux and double-flux detection avoidance enable a given domain name to rapidly modify the IP addresses associated with the attacker's domain name servers and hosts. But if resolution of the example.com domain itself can be prevented, the C&C center could be effectively shut down. As we will discuss in more detail later in this chapter, DNS firewalling can be deployed to block or otherwise modify query responses for given domains and DNS server names or IP addresses, among other criteria.

Dynamic Domain Generation

In the ever-escalating arms race between attack threats and attack defenses, the tactic of dynamically generating domain names themselves provides a moving target on the query name as well. Domain generation algorithms (DGAs) can be used by malware to dynamically generate a set of candidate domains periodically to reach the C&C center. The botnet operator likewise uses the DGA to calculate a domain name which they can register such that the DGA domain is resolvable at the time when the bots attempt to query it to enable communications. This technique was used by CryptoLocker ransomware and the family of Conficker worms, for example, which generated from 250 to 50,000 domain names per day in its varying forms. This third form of flux with dynamic domain names on top of dynamic name server IP addresses and resolution IP addresses enable malware to better elude detection as a continually moving target.

DETECTING MALWARE

Up-to-date anti-malware software installed on all network attached devices is recommended as a first line of detection at the device level to reject malware installation attempts. Frequent changing of passwords is also recommended as is keeping kernel, operating system, and application software patched and updated. Some firewalls and intrusion prevention systems can also be configured to block malware infiltration based on traffic patterns or signatures. Changing of vendor default user IDs and passwords, along with other host controls are necessary as well.

DNS itself can be used to block certain malware communications and identify infected hosts as we will discuss later in this chapter. But as malware producers develop new "strains" to outwit defenses, anti-malware vendors characterize behavior then develop remedies to quarantine or extricate the malware. Malware producers then seek new methods and the arms race spirals onward.

Detecting Malware Using DNS Data

While DNS firewalls can block malware attempts to contact a known C&C center, the effectiveness of the firewall is only as good as the quality of its blocklist, much as the effectiveness of anti-malware host software depends on the quality (which often maps to timeliness) of its host detection technology. Much as anti-malware software vendors remain vigilant for the presence of new installations or files to identify potential malware, DNS query data can likewise be analyzed. Malware escaping host detection can possibly be discovered based on the DNS queries it issues to locate its C&C center.

By analyzing your DNS query data, you may be able to identify the presence of malware within your network before your antivirus software has remedied a block for it. While relying exclusively on DNS query data to detect malware is by no means recommended, DNS query analytics can serve as valuable input to your overall malware detection "eco-system" and mitigation strategy. Data from hosts, routers, firewalls, intrusion detection systems, and DNS can all help identify suspicious and malicious activity.

In this section, we'll discuss characteristics of DNS query traffic that may signal the existence of malware attempting to utilize DNS to locate its C&C center or for use in communications. Note that none of these characteristics are certain indicators of the presence of malware. In the interest of minimizing false positives, you face walking the tightrope of minimizing the blocking of valid queries versus enabling malware access to its C&C for software and/or instructions. The presence of multiple characteristics or repeated presence of one or more over a brief period of time may increase the probability of malware's existence, but decision policies should be defined and tweaked over time to arrive at an acceptable tradeoff.

Fast-Flux Networks   Malware traffic identified by routers, firewalls, or intrusion detection or prevention systems can typically be blocked by corresponding gateways. However, if the attacker cycles through a variety of IP addresses for its C&C center, blocking by destination IP address will be ineffective. Even blocking by domain name may likewise not work for long. Malware developers have evolved (and will continue to evolve) techniques to communicate between infected hosts and the C&C center while evading detection and take down.

Key characteristics of fast-flux networks include the following. Scoring domain names along these attributes could enable characterization of likelihood of a domain being of a fast fluxing nature.

- Low TTL (of the order of 300 s)
- Large number of resolved IP addresses for the domain over time

- High rate of change of resolved IP addresses for the domain over time
- Resolved IP addresses for the domains are scattered across several different networks as indicated by association with multiple autonomous system numbers (ASNs), network names, and organization names over time
- Resolved IP address from a different country than that implied by a domain beneath the corresponding country-code TLD
- Short domain lifetime
- Bursty resolution activity for the domain
- Reputation qualification of domain registrar and ASN

Notice that some of these characteristics are shared by legitimate content delivery networks (CDNs) so care must be taken to avoid identifying such as false positives. Several approaches for detecting fast-flux networks have been proposed within the research community. Some approaches involve monitoring in-band traffic characteristics such as HTTP transaction delays or correlating incoming and outgoing TCP connections within a stub (e.g., residential broadband) network. We won't discuss these here but will focus on those that rely on DNS query data to detect fast-flux networks. Two basic forms of detection have been proposed.

ACTIVE DETECTION   An *active* approach entails the compilation of domain names as captured by spam archives followed by the repeated querying of these domains to identify TTL values, resolved IP addresses, and networks from which resolved IP addresses are sourced. Based on queried data, a decision is made regarding the likelihood of a given domain being of a fast-flux nature. The active approach can support faster detection of malware queries, though it suffers from being based on an already defined source set of potentially fluxing domains and recurring querying may tip off fast-flux network operators of detection attempts and impel them to "go dark" for a time to evade detection. Also, active probing does not inherently enable capture of other fast-flux characteristics such as domain lifetime and domain-IP address history.

PASSIVE DNS   A *passive* approach leverages DNS query data collected over a broader sample set (e.g., the Internet) over a longer time period. *Passive DNS replication* was originally proposed by Florian Weimer "to obtain domain name system data from production networks, and store it in a database for later reference" (75). The passive DNS architecture consists of sensors which capture DNS packets. The sensors are placed on the Internet-facing network between caching servers and the Internet; for example, within the DMZ. This enables the collection of queries for which cached data does not exist, cutting down packet volume and also serves to anonymize queries as originating from the recursive servers instead of individual clients. This is beneficial if you desire to deploy sensors to provide input to a passive DNS aggregator, which provides the value of a broader DNS query data set based on multiple sensors' inputs.

Several passive DNS providers including TotalPassive, Circl, and Farsight Security accept feeds from deployed sensors to collect DNS query traffic for storage within each provider's centralized repository. Subscribers may then query the repository for Internet-wide DNS data.

Such information can be instructive in helping to answer DNS questions such as

- What domain names are hosted by a given DNS server?
- What domain names resolve address information to a given IP address or network?
- What IP address(es) did a given domain name resolve to in the past?
- What subdomains exist beneath a given domain in the hierarchy?
- How recently was a given domain "seen"?

Answers to such questions can aid in an investigation of a new or suspicious domain names, DNS servers, or suspect IP address blocks. Considering the history of a given domain name, DNS server name or IP address, or ASN from which its IP address originates can provide a reputation assessment for a given domain name. Published research leveraging passive DNS data include FluxBuster (76), Fast-flux hunter (77), that proposed in (78), and several others.

Both active and passive detection approaches can be useful to provide diverse DNS data to a centralized security information and event management (SIEM) systems, where active monitoring can provide early warning while passive data can corroborate or dismiss the active detection. Both data sets could be useful for historical and forensic analysis.

Domain Generation Algorithms (DGAs)   DGAs enable the botnet controller to supply an algorithm with its installed malware such that the malware can calculate a set of possible domain names to which to issue queries to locate the C&C center. The botnet operator utilizes the same algorithm to publish the corresponding domain name for DNS rendezvous at the corresponding time. Many such domains are activated for 1 day or even just a few hours to communicate with bots and to obviate the effectiveness of domain shutdown.

Installed malware may issue dozens of queries a day to seemingly random domain names. Candidate domain names are generated based on a salt parameter and a random seed, sometimes the current date/time, fed into an iterative algorithm. The output is a unique string of random characters which is used as a domain label in a query for the C&C center. Until the C&C center is published in DNS with corresponding resolution data, a relatively large number NXDOMAINs will likely be received.

Early forms of detection focused on the "randomness" of domain labels as differing from human created domain names. However, these days many advertising links, CDNs and other automatically generated domain names likewise utilize algo-rithms that may match the profile of a randomized domain label. The observation of a larger than normal set of NXDOMAINs may indicate attempts to contact a DGA-named C&C center (or a bogus queries attack). An example detection implementa-tion was posited as a system called Pleiades (79) and others have been proposed since then. Like fast fluxing, DGA detection research continues and commercial domain blocking vendors are striving to maintain pace to provide customers updated block lists.

MITIGATING MALWARE USING DNS

Malware Extrication

If installed anti-malware software has not yet been updated to identify and protect against the strain of malware attacking your systems, it is likely a new form or variant. Such attacks are referred to as *zero-day attacks* in that the target has zero days to prepare defenses and mitigation tactics. Identify infected systems by tracing detected events to corresponding systems utilizing the associated IP address at the time of the attack. If possible, quarantine the device(s) from your production network to help prevent further spreading and reimage infected devices and apply backed up data to restore the device to the extent possible.

Review network and DNS logs to identify suspicious activity such as unusual DNS traffic that contains client queries with a higher percentage of NXDOMAIN responses or large query payloads. Research queried domains from the general attack time period to try to identify potential malware C&C centers to block. Suspicious domains may be those that are new or short-lived based on fluxing or DGA. You can subscribe to passive DNS services to enable lookups for such information.

The DNS servers that responded authoritatively to queries for suspicious domains may also provide input to building a case. DNS server operators have var-ious levels of security controls in place and certain operators may be less reputable in terms of screening domain administrators. Thus, the domain names themselves of responding DNS servers may provide clues as well as the IP addresses of these servers.

However, be careful when considering blocking IP addresses as a single IP address could be shared by several servers and may impose unacceptable collateral damage in blocking reputable servers. Nevertheless, such an initiative may be war-ranted at least during the time period during which the malware mitigation and attack recovery phases are completed.

It's important to further broader industry awareness of new attacks and attempted defense tactics, successful or otherwise. Together we make the Internet more secure. Report what information is available to your anti-malware vendor, to relevant industry groups, and/or to the US-CERT Malware Analysis site.

DNS Firewall

As we saw in Chapter 5, the configuration of a recursive DNS server is generally straightforward and consists of a hints file containing a list of root servers from which the domain tree traversal may begin when seeking query answers and perhaps additional ACLs and cache management parameters. The security configuration is more involved as we shall see. Given the recursive server serves as each client's gateway to the global DNS, it not only serves as the point from which query answers are derived but as the last point from which query answers from internal or external networks are returned to those clients.

With this in mind, the ISC BIND, PowerDNS, and Knot DNS implementations provide the ability to modify received query answers or downright deny certain query answers in the interest of security or governance policies. This ability to deny or modify DNS protocol level transactions has led to the term, "DNS firewall." Unbound has stated plans to develop DNS firewall functionality in 2017.

A DNS firewall enables you to configure DNS to modify answers to certain queries or queries answered from particular servers. While configuring DNS to "lie" was initially frowned upon by Internet purists, unfortunately the nature of the Internet today necessitates consideration of passing on query answers that can lead to data exfiltration and/or manipulation of a given host or others for nefarious purposes. Other forms of query answer manipulation can also be configured with recursive servers, such as DNS64 for IPv4 reachability from IPv6 networks and NXDOMAIN redirection to search pages for monetization.

DNS firewall technology provides the ability to block certain DNS responses based on queries to known bad actor domains, among other criteria. It also provides for the modification of DNS responses to clients to redirect clients to an internal "walled garden," for example. DNS firewall technology enables enterprises and service providers to reduce the probability of malware proliferation within their networks. It also enables service providers to offer incremental revenue features such as parental controls and related domain blocking, while enabling compliance with domain or end user blocks imposed by law enforcement.

A DNS firewall is also known as Response Policy Zones (RPZs), which is the DNS implementation term that provides the DNS firewall functionality. An RPZ is a special zone that defines the triggers and associated query response policies for queries or answers matching each trigger.

To view how this works, consider Figure 11.3. Malware installed on the client laptop on the left of the figure issues a query to locate its C&C center at bad-example.com. The recursive server processes the query normally, traversing down the domain tree in the absence of cached information to obtain an authoritative answer from the bad-example.com DNS server. However, our DNS administrator has configured an RPZ zone which is intended to direct the user to a captive portal.

In the scenario above, the recursive server checks its RPZ based on the answer received which resolved the question to 192.0.2.254. A response policy is defined

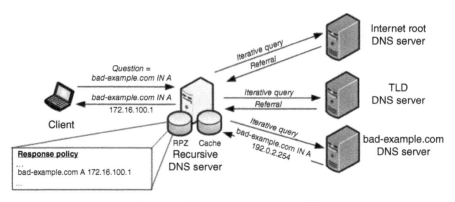

Figure 11.3. DNS Firewall Example

in this case for query answers where the Qname was bad-example.com. This match instructs the recursive server to modify the response provided to the client based on the policy, which in this case features a resolution redirect to 172.16.100.1, our captive portal server.

RPZ feeds can be generated and managed internally and/or by leveraging external RPZ data providers such as Farsight Security, DissectCyber, SpamHaus, SURBL, Diamondback Networks, and ThreatStop. RPZ information defines various policy triggers and corresponding policies for query disposition. Triggers can be defined as:

- Qname match – the queried domain name (or within a domain branch) matches the owner field of the RPZ resource record. For example, an RPZ resource record with owner name good.example.com will apply the policy to responses to queries with the matching Qname.
- RPZ-IP – query answer IP address (or block) matches the IP address in the general format: <prefix length>.<reversed IP address>.rpz-ip. For example, to match an answer with an IP address from the 192.0.2.0/25 block, this would be encoded as 25.0.2.0.192.rpz-ip. An IPv6 address can be similarly encoded with the abbreviation of "zz" for a double colon abbreviation; for example, one could match an IPv6 address in the answer from the 2001:db8:11b::1025/128 as 128.1025.zz.11b.db8.2001.rpz-ip.
- RPZ-NSIP – authoritative name server IP address, encoded similarly as rpz-ip matches but with the rpz.nsip suffix.
- RPZ-NSDNAME – authoritative name server domain name, for example, ns1.bad-example.com.rpz-nsdname.
- RPZ-CLIENT-IP – client IP address formatted as above with prefix length and reversed IP address and with the rpz-client-ip suffix.

Triggers are matched on a best fit basis. That is all records with triggers matching the query or answer are sorted by the best fit (e.g., longest prefix match for IP addresses) or most granular match; the policy corresponding to such a best match trigger is then applied. Based on particular triggers, policies can be defined to answer the query with either

- NXDOMAIN – respond to the querier that the queried name was not found; this policy is denoted with a CNAME RRType and RData field containing. (i.e., dot).
- NODATA – signified by a NOERROR response with no query answers (a zero "answer count"); this indicates a queried name was valid but no data for the queried type was found. This policy is denoted with a CNAME RRType and RData field containing ∗. (i.e., asterisk dot).
- Pass-through – no alteration of the answer, and is denoted with a CNAME RRType and RData of `rpz-passthru`.
- Drop – drops the query and provides no answer; denoted by a CNAME RRType with RData of `rpz-drop`.
- TCP-Only – this policy indicates a short response with the truncated header bit set to instill the client to retry using TCP. This is similar to the response rate limiting slip functionality to mitigate distribute denial of service reflector attacks. This policy is indicated with a CNAME RRType with RData of `rpz-tcp-only`.
- Local policy such as directing the querier to portal or walled garden web page indicating an invalid query or otherwise communicating with the user. This policy can be specified as a "regular" RRType (A, AAAA, CNAME) with RData containing the corresponding IP address or domain to provide as the answer to the query.

Consider the following RPZ file for examples of how to define policies for defined triggers.

```
$TTL 1h
@      SOA   localhost admin.example.com (1 1h 15m 28d 2h)
       NS         localhost.

; Qname trigger examples (no dots after owner field)
badanswer.example.com    CNAME   .    ;NXDOMAIN policy
*.badanswer.example.com CNAME    .    ;apply to subdomains

; IP address in answer trigger
; In this example we answer "no data" when the query answer
; is within 192.0.2.0/24 except we pass through 192.0.2.5
; due to its match of the 32 mask length as best match
```

```
24.0.2.0.192.rpz-ip       CNAME   *.              ;NODATA policy
32.5.2.0.192.rpz-ip       CNAME   rpz-passthru. ;pass through

; drop query answers from these name servers:
;      ns.hack.example.com and IP address 2001:db8::1f72
ns.hack.example.com.rpz-nsdname  CNAME     rpz-drop.
128.1f72.zz.db8.2001.rpz-nsip    CNAME     rpz-drop.

;force querier 192.0.2.1 to use tcp
32.1.2.0.192.rpz-client-ip   CNAME   rpz-tcp-only.

;redirect to web portal (local data)
wrongway.example.com    A    192.168.0.1 ; Qname trigger
```

It's important that you configure your DNS server to log RPZ data hits. That is log each occurrence of the triggering and application of a response policy. The source IP address should prove indispensable in identifying the device issuing the query which is potentially malware infected. If your DNS firewall is configured on recursive servers directly queried by stub resolvers, logging should indicate the culprit directly. However, if you've deployed a tiered recursive DNS server architecture with DNS firewalling configured on your Internet caching servers, logs from these servers will indicate the source of the query as the originating recursive server. Further investigation of the corresponding recursive server logs will be necessary to attempt to identify the end device that had originated the query.

SUMMARY

In addition to DNS firewall configuration, recursive servers may also be configured to perform DNSSEC validation to enable validation of signed resource record sets. This requires configuration with the current Internet root zone public key (DNSKEY KSK record data). Caching servers can also be configured to automatically maintain this trust anchor public key as the root zone rolls.

Other recursive security settings should also be configured to provide maximum availability and reliability to the DNS recursive service. Such settings include setting ACLs to constrain from what IP networks or addresses recursive queries may be made, for example, configuring denial of service rate limiting if available, restricting the number of outstanding queries per client and limiting the query depth to minimize the chasing of bogus queries. These settings are examples and your name server implementation may offer different and/or additional security features as well.

A DNS firewall is only as good as the triggers and policies with which you configure it. Organizations such as spamhaus, surbl, and commercial organizations such as Diamondback Networks, ThreatStop and ThreatConnect can provide RPZ data feeds directly to your recursive servers to keep this block data relatively fresh. You should consider their data sources and configuration policies to maximize blocking of malware communications while minimizing false positives. When assessing whether a domain should be blocked or not, consideration of its history could provide valuable input; passive DNS is one source of such information.

12

DNS SECURITY STRATEGY

This chapter summarizes key elements of a DNS security strategy. Available, responsive, and accurate DNS service is an absolute must in today's IP networks. A multifaceted defense in depth approach is required to provide a secure defense against attackers, natural disasters, and human errors. We'll summarize this first with respect to basic approaches to mitigate major vulnerability types we discussed in this book. Then we'll summarize tactics for inclusion in your overall security strategy, securing your infrastructure, and specific strategies for each major DNS role or function. Finally, we'll summarize other Cybersecurity Framework outcomes you should consider when developing a comprehensive security management plan.

The diversity of DNS threats presents a serious challenge to IT administrators responsible for managing the operation and integrity of DNS services within their networks. With such a variety of attacks, no single mitigation technology or approach will suffice in defending against them all. A mitigation portfolio is necessary to provision layered defenses for each major attack threat. The preceding chapters defined techniques available to DNS administrators in assembling such a mitigation strategy for each of these vulnerabilities and risks summarized in this chapter.

DNS Security Management, First Edition. Michael Dooley and Timothy Rooney.
© 2017 by The Institute of Electrical and Electronic Engineers, Inc. Published 2017 by John Wiley & Sons, Inc.

MAJOR DNS THREATS AND MITIGATION APPROACHES

As we've seen, several varieties of DNS attacks are possible to disrupt DNS or network communications in general or to leverage DNS' intended purpose to identify targets or attack systems with malicious intent. No single mitigation approach can eliminate vulnerabilities to all threats; thus, a multipronged mitigation strategy is required to reduce attack exposure. Table 12.1 summarizes the threats to DNS and corresponding mitigation approaches for each.

COMMON CONTROLS

DNS servers, like all network infrastructure must be protected through the implementation of "common controls," applicable to all core infrastructure components to protect these resources from malicious attack, illicit use by an attacker, inadvertent errors, and natural disasters.

Disaster Defense

Predicting natural disasters such as earthquakes, hurricanes, floods is imperfect though the likelihood of such events can normally be estimated based on geographic deployment of infrastructure. Nevertheless, even if you've deployed in an area which has yet to experience severe weather or cataclysmic events, you're never fully immune to such events in the future. Certainly, in this day and age, unnatural disasters are also possible anywhere. Such disasters might include terrorist attacks or acts of war against you or against a primary supplier, for example, your power company. This of course is all part of the likelihood component that you and your organization needs to assess when assessing such risks.

Defending yourself against disasters, natural or otherwise, generally requires deployment of your infrastructure in two or more locations that are at least not prone to the same set of natural disasters. Deploying more than two sites is better but obviously costlier. But with two sites, each housing critical infrastructure sufficient to support the organization on its own, you should be able to withstand the scenario when one site is rendered unavailable.

From a DNS perspective, elements of each trust sector should be deployed in each major site unless you have deployed more than two or three of each elsewhere. Thus, you may deploy a set of Internet caching servers and a set of external authoritative servers in each major site but a set of internal authoritative or extranet authoritative DNS servers in remote offices. The general guideline is to deploy a DNS server within each trust sector you require in at least two diverse sites.

Consider your network and security management systems as well, which must also survive a major site outage. Deployment of a replicated backup system for each can provide critical information and controls when you most need it.

TABLE 12.1 DNS Threats and Mitigation Approach Summary

	Threat	Threat Summary	Mitigation Approaches
Denial of Service	Denial of service	Attacker transmits a high volume of TCP, UDP, DNS, or other packets to the DNS server to inundate its resources	• Inbound port and DNS ACLs • Inbound rate limiting • Limit TCP connections • Anycast deployment • Load balancers • Use multiple servers in-house and service provider • DNS cookies
	Distributed denial of service	Attacker transmits a high volume of TCP, UDP, DNS, or other packets from multiple sources to the DNS server to inundate its resources	• Inbound port and DNS ACLs • Inbound rate limiting • Limit TCP connections • Anycast deployment • Load balancers • Use multiple servers in-house and service provider • DNS cookies
	Bogus queries	Attacker transmits a high volume of bogus queries, causing the recursive server to futilely locate authoritative answers	• Limit the number of outstanding queries per client • Configure jostle timeout or equivalent for throttling • Limit query depth
	PRSD	Attacker transmits a high volume of bogus queries with a common domain suffix to impact recursive and authoritative servers	• Limit the number of outstanding queries per client • Configure jostle timeout or equivalent for throttling • Apply fetch limits

(continued)

TABLE 12.1 *(Continued)*

	Threat	Threat Summary	Mitigation Approaches
Cache Poisoning	Packet interception/spoofing	Attacker transmits a DNS response to a recursive DNS server in order to poison its cache, affecting DNS resolution integrity	• DNSSEC validation on recursive servers • Source port and TXID randomization • Qname case manipulation and verification on response
	ID guessing/query prediction	Attacker transmits a DNS response(s) to a predicted query using a predicted or variety of TXID values	• DNSSEC validation on recursive servers • Source port and TXID randomization • Qname case manipulation and verification on response
	Kaminsky attack/name chaining	Attacker transmits a DNS response(s) with falsified answers in the DNS message Additional section. The Kaminsky attack produces deterministic queries to facilitate the attack	• Source port and TXID randomization • Use updated DNS software versions with bailiwick and other checks • DNSSEC validation on recursive servers
Authoritative Poisoning	Illicit dynamic update	Attacker transmits a DNS Update message(s) to a master DNS server to add, modify, or delete a resource record in the target zone	• Use ACLs on allow-update, allow-notify notify-source • Define update policies to constrain scope of allowed updates, e.g., by RRType • Define ACLs to require transaction signatures for origin authentication and data integrity

TABLE 12.1 (*Continued*)

	Threat	Threat Summary	Mitigation Approaches
Authoritative Poisoning	Server attack/hijack	Attacker hacks into the DNS server which enables manipulation of DNS data among other server capabilities	• Harden your operating system • Manage host access controls • Use hidden masters to inhibit detection of the zone master • Monitor vulnerability sites and keep software updated • Limit port or console access
	DNS service misconfiguration	Vulnerability to configuration errors exposes the DNS service to improper configuration	• Use checkzone and checkconf or similar utilities • Use an IPAM system with error checking • Keep fresh configuration backups for reload if needed • Monitor your DNS on service provider servers for anomalies
	Domain hijacking	An attacker hacks your domain registrar account or your email to change your domain information	• Choose registrars that provide a layered authentication system • Monitor for changes in your domain information • Secure your email account, frequently changing passwords

(*continued*)

TABLE 12.1   *(Continued)*

	Threat	Threat Summary	Mitigation Approaches
Server/OS Attack	Buffer overflows and OS level attacks	Attacker exploits server operating system vulnerability	• Harden operating system • Monitor vulnerability sites and keep software updated
	Control channel attack	Attacker accesses the DNS service control channel to disrupt DNS service	• Control channel ACLs • Control channel keyed authentication • Constrain scope of control channel functions
	DNS service vulnerabilities	Attacker exploits DNS service vulnerability	• Monitor CERT advisories and update DNS software • Do not expose DNS service version to version queries
Resolver/Host Attacks	Recursive DNS redirection	Attacker misconfigures resolver to point to illicit recursive DNS server	• Configure DNS servers via DHCP • Monitor for rogue DHCP servers • Periodically audit each client for misconfigurations or anomalies
	Resolver configuration attack	Attacker hacks into the device which enables manipulation of resolver configuration among other device capabilities	• Manage host access controls
Network Reconnaissance	Name guessing	Attacker issues legitimate DNS queries for names that, if resolved could serve as further attack target	• Avoid naming hosts with overly "attractive" names • Define ACLs on queries
	Illicit zone transfer	Attacker initiates a zone transfer request to an authoritative DNS server to obtain zone resource records to identify potential attack targets	• Use ACLs with TSIG on allow-transfer; and use transfer-source IP address and port to use a nonstandard port for zone transfers

TABLE 12.1   (*Continued*)

	Threat	Threat Summary	Mitigation Approaches
Reflector Style Attacks	Reflector attacks	Attacker spoofs the target's IP address and issues numerous queries to one or more authoritative DNS servers to inundate the target	• Implement ingress filtering on routers to mitigate spoofing • Use DNS response rate limiting • Consider limiting ANY query answer content
	Amplification attacks	Attacker amplifies reflector attack by querying for "large" resource records to increase data flow to target per query	• Implement ingress filtering on routers to mitigate spoofing • Use DNS response rate limiting • Consider limiting ANY query answer content
Data Exfiltration	DNS tunneling	Attacker transmits data through firewalls using DNS as the transport protocol	• Monitor DNS queries for frequent queries between a given client and server especially with large query and response payload
	Resource locator	Attacker infects internal device which uses DNS to locate command and control center	• DNS firewall • Use passive DNS to investigate suspect domains or IP addresses
APT	Advanced persistent threats	Attacker deploys adaptable malware within a network to perform nefarious functions to disrupt communications and/or steal information	• DNS firewall • Use passive DNS to investigate suspect domains or IP addresses

Defenses Against Human Error

Human errors happen all the time and in ways often not contemplated. Errors may include misconfigurations of DNS yielding inaccurate resolution, networking errors hampering optimal routing and resolution performance, and users unwittingly downloading malware, among others. You can protect your network through network and network management controls, but periodic user training and communications is also critical to increasing security awareness and reducing errors.

User roles and responsibilities should be clearly enumerated and understood by each user based on the user's particular job role. General security practices related to physical security, password changing, and safe browsing, and emailing should be conveyed as new employees come on board and as part of regular required training programs.

Convening change control meetings among relevant stakeholders enables communication of proposed network and system changes and affords the opportunity to others affected to voice objections or concerns regarding the change scope or timing. A back out strategy should also be presented in the event of unforeseen issues. Gaining consensus is key to getting everyone on the same page and moving forward with planned changes with requisite support on board should changes yield unanticipated or unsuccessful results.

Use of not only disciplined processes but management systems can also reduce human errors. An IP address management (IPAM) system can help identify and correct potential configuration errors before they are introduced into the production network. Such systems also provide audit logging to enable review of which administrator performed which task for accountability and troubleshooting purposes.

DNS ROLE-SPECIFIC DEFENSES

This section summarizes the defenses by DNS component for convenience.

Stub Resolvers

The following tactics should be implemented for your stub resolvers:

- Host controls including physical, operating systems, and resolver software
- DHCP server audits, for example, LeaseQuery to confirm IP addresses on network corresponding to DHCP leases

- DNS cookies
- Resolver connection encryption such as DNSCrypt

Forwarder DNS Servers

The following tactics should be considered for implementation on your forwarder DNS servers:

- Plan deployment within recursive trust sector – size number and capacity of servers
- Host controls including physical, hardening, operating systems, and DNS software
- Server network interface ACLs
- DNS software "allow-query" ACLs
- Limit queries per client
- DNS cookies
- Client connection encryption such as DNSCrypt

Recursive Servers

The following tactics should be considered for implementation on your recursive DNS servers:

- Plan deployment within recursive trust sector – size number and capacity of servers
- Anycast addressing
- Host controls including physical, hardening, operating systems, and DNS software
- Server network interface ACLs
- DNS software "allow-query" ACLs
- Source port and transaction ID randomization
- Query case randomization
- DNSSEC validation
- DNS firewall
- Fetch query or jostle timeout controls
- Limit queries per client
- Query log audits for tunneling, malware connections
- DNS cookies
- Client connection encryption such as DNSCrypt

Authoritative Servers

The following tactics should be considered for implementation on your internal, external, and extranet authoritative DNS servers:

- Plan deployment within appropriate authoritative trust sector (external, extranet, or internal namespace) – size number and capacity of servers
- External DNS service provider backup or diversity
- Host controls including physical, hardening, operating systems, and DNS software
- Disable recursion
- Anycast addressing
- Sign zones with DNSSEC
- Restrict zone updates to specific servers, users, IPAM system, and so on
- Restrict zone transfers to specific servers requiring signatures
- Query ACLs if appropriate based on deployment
  - Internal authoritative servers – allow query for internal clients only
  - Extranet authoritative servers – allow query for partner clients if possible
  - External authoritative servers – not generally scoped by nature as deployment as Internet queryable resources
- Response rate limiting

BROADER SECURITY STRATEGY

The DNS security controls we've discussed in this book related primarily to host and data protection for DNS servers and protocol, respectively. However, security risk analysis and mitigation also necessitates education and enforcement for your end users and IT staff. Security needs to be ingrained into the culture of your organization. The Cybersecurity Framework Core encompasses this critical human aspect of security management and the following controls should be applied to DNS-specific risk analysis and controls and incorporated into your overall network security strategy. This section summarizes these broader security aspects.

Security knowledge and awareness must pervade all levels of the organization. This is not to say that every employee, partner, or associate must memorize your full security management plan. But you should carve out relevant sections of the plan to convey to each stakeholder with respect to security awareness and suspicious activity reporting.

The following cybersecurity framework subcategories provide high level guidance on broader organizational considerations beyond DNS-specific controls. Please consult Appendix A for a listing of all subcategories with example DNS-specific desired outcomes, and refer to the Cybersecurity Framework document and supporting documents for full details.

Identify Function

Asset and access management subcategory items of the Identify category deal primarily with deployment and server controls discussed in Chapters 5 and 6, respectively, while risk management items apply to all DNS-specific threats covered in the prior seven chapters. Most of the other subcategory items are summarized below.

- **ID.AM-3** – Organizational communication and data flows are mapped – Map the flow of information from event detection, response, and recovery including communications to relevant stakeholders at various points in the process.
- **ID.AM-6** – Cybersecurity roles and responsibilities for the entire workforce and third-party stakeholders are established – Include security elements within periodic employee training at all levels of the organization including what to look for and who to call for reporting possible events.
- **ID.BE-1** – The organization's role in the supply chain is identified and communicated – Incorporate security requirements and features when procuring goods from suppliers and define and communicate contingencies for supply shortages, global events, and other potential supply-impacting events.
- **ID.BE-2** – The organization's place in critical infrastructure and its industry sector is identified and communicated – Security requirements are included and communicated with respect to critical organizational infrastructure planning and management.
- **ID.BE-3** – Priorities for organizational mission, objectives, and activities are established and communicated – Security is feathered into all aspects of the corporate mission, objectives, and activities and communicated to all stakeholders.
- **ID.BE-4** – Dependencies and critical functions for delivery of critical services are established – Incorporate security into business process definitions and deploy redundantly for critical network services.
- **ID.BE-5** – Resilience requirements to support delivery of critical services are established – Consider security requirements in procurement decisions including redundancy, high availability, and security features.
- **ID.GV-1** – Organizational information security policy is established – Publish an information security policy document which is approved by management and published to stakeholders.
- **ID.GV-2** – Information security roles and responsibilities are coordinated and aligned with internal roles and external partners – Roles and responsibilities for security functions and processes are documented, communicated to internal and relevant external associates.

- **ID.GV-3** – Legal and regulatory requirements regarding cybersecurity, including privacy and civil liberties obligations, are understood and managed – Include relevant legal and regulatory impacts in your information security policy document which is approved by management and published to stakeholders.
- **ID.GV-4** – Governance and risk management processes address cybersecurity risks – Document the risk management process and have it approved by management and communicated to employees and external stakeholders.
- **ID.RA-6** – Risk responses are identified and prioritized – The organization's cumulative list of threats, vulnerabilities, likelihoods, and business impacts are documented and prioritized with respect to defense prioritization.
- **ID.RM-1** – Risk management processes are established, managed, and agreed to by organizational stakeholders – Establish risk management processes including the identification, classification, and business impact assessment for identified risks. Garner agreement by stakeholders.
- **ID.RM-2** – Organizational risk tolerance is determined and clearly expressed – Define and gain consensus regarding the process for determining and documenting organizational risks.
- **ID.RM-3** – The organization's determination of risk tolerance is informed by its role in critical infrastructure and sector-specific risk analysis – Define and gain consensus regarding the assessment of risk tolerance for each identified risk.

Protect Function

Protection strategies were discussed within the prior seven chapters for corresponding vulnerabilities and risks. These items below relate across these and all components of network security.

- **PR.AT-1** – All users are informed and trained – Devise, develop, and deliver training for IT security. Keep training material updated and provide periodic training updates.
- **PR.AT-2** – Privileged users understand roles and responsibilities – Document well-defined job descriptions outlining roles and responsibilities including security aspects and garner acknowledgement from associated users.
- **PR.AT-3** –Third-party stakeholders (e.g., suppliers, customers, partners) understand roles and responsibilities – Obtain acknowledgement from third-party stakeholders of security related roles and responsibilities.
- **PR.AT-4** – Senior executive users understand roles and responsibilities – Provide training and garner acknowledgment of roles and responsibilities for executives

- **PR.AT-5** – Physical and information security personnel understand roles and responsibilities – Provide training and garner acknowledgement of roles and responsibilities for physical and information security personnel.
- **PR-IP-2** – A Systems Development Lifecycle to manage systems is implemented – Apply security engineering principles to your IT network and systems, including in-house development projects and product procurement analysis.
- **PR.IP-6** – Data is destroyed according to policy – Written and computing device data deemed for disposal is destroyed according to policy to prevent information theft.
- **PR.IP-7** – Protection processes are continuously improved – Data protection processes are continuously improved through periodic review, new technologies, and lessons learned.
- **PR.IP-8** – Effectiveness of protection technologies is shared with appropriate parties – DNS data protection technology effectiveness is shared with appropriate parties.
- **PR.IP-9** – Response plans (Incident Response and Business Continuity) and recovery plans (Incident Recovery and Disaster Recovery) are in place and managed – Incident response and recovery plans are documented, communicated and managed, as are business continuity and disaster recovery plans.
- **PR.IP-10** – Response and recovery plans are tested – Incident response and recovery plans are tested and results fed back to improving such plans.
- **PR.IP-11** – Cybersecurity is included in human resources practices (e.g., deprovisioning, personnel screening) – Security is incorporated into human resources processes relating to recruiting, hiring, evaluations as appropriate, training, and deprovisioning.
- **PR.IP-12** – A vulnerability management plan is developed and implemented – A vulnerability management plan describing the process for vulnerability identification, assessment, likelihood analysis and potential impacts is developed and implemented.

Detect Function

Tactics for detecting attacks or risk instances were discussed within the prior seven chapters for respective attack vectors. The items below address the detection processes category which applies across these and all components of network security.

- **DE.DP-1** – Roles and responsibilities for detection are well defined to ensure accountability – Personnel roles and responsibilities for network and DNS security event detection within the organization are well defined.

- **DE.DP-2** – Detection activities comply with all applicable requirements – Security event detection activities are documented and enforced in accordance with event detection requirements.
- **DE.DP-3** – Detection processes are tested – Security event detection activities and systems are tested to characterize detection effectiveness.
- **DE.DP-4** – Event detection information is communicated to appropriate parties – Security event detection information is communicated to appropriate parties in accordance with the incident response plan.
- **DE.DP-5** – Detection processes are continuously improved – Security event detection processes are continuously improved based on technology or process improvements as well as lessons learned from prior events.

Respond Function

We discussed vulnerability specific analysis and mitigation tactics in prior respective chapters. However, it's recommended within the cybersecurity framework to amalgamate information across various inputs to help fully classify attack breadth and impact; hence, we summarize these here.

- **RS.RP-1** – Response plan is executed during or after an event – As DNS security events are detected and characterized, relevant actions from the incident response plan are executed.
- **RS.CO-1** – Personnel know their roles and order of operations when a response is needed – Personnel roles and responsibilities for DNS security event response within the organization are well defined within the incident response plan.
- **RS.CO-2** – Events are reported consistent with established criteria – DNS security events are detected and characterized in accordance with established criteria.
- **RS.CO-3** – Information is shared consistent with response plans – DNS security event response information is communicated to appropriate parties in accordance with the incident response plan.
- **RS.CO-4** – Coordination with stakeholders occurs consistent with response plans – DNS security event response information is communicated to appropriate stakeholders in accordance with the incident response plan.
- **RS.CO-5** – Voluntary information sharing occurs with external stakeholders to achieve broader cybersecurity situational awareness – DNS security event response information is communicated to external stakeholders in accordance with the incident response plan to facilitate industry awareness of the attack and effective defensive measures.

- **RS.AN-1** – Notifications from detection systems are investigated – DNS event detection systems are investigated to characterize the event as a potential attack or threat.
- **RS.AN-2** – The impact of the incident is understood – Upon incident detection, the incident should be analyzed to assess and understand the impact of the incident. The incident response plan should be followed and impacted groups involved in responding to contain, eradicate, and recover from the incident, while communicating status in accordance with the response plan. New information or lessons learned should be incorporated into an update of the response plan based on review, concurrence, and approval by appropriate members of the organization.
- **RS.AN-3** – Forensics are performed – Forensic analysis on detected incidents should be performed to go beyond the symptoms of the incident to identify the ultimate cause and to enumerate those vulnerabilities exploited or attacked. This analysis is useful for identifying new or morphed attack vectors and vulnerabilities, and to qualify the effectiveness of any defensive controls that were intended to protect against such an attack.
- **RS.AN-4** – Incidents are categorized consistent with response plans – Incidents need to be categorized in a manner consistent with incident response plans. This is helpful in terms of prioritizing actions and inclusion of appropriate staff to analyze, contain, eradicate, and resolve the incident in a timely manner.
- **RS.IM-1** – Response plans incorporate lessons learned – After incident recovery, a postmortem discussion with involved staff is useful for reviewing the incident, possible defensive and mitigation steps to improve response and recommended response plan updates to incorporate lessons learned.
- **RS.IM-2** – Response strategies are updated – Incident response strategies should be reviewed and updated as appropriate.

Recover Function

Much like the respond function, where a more complete and effective response can be applied across DNS servers and related networks and servers, the recover function broadly applies to enable full restoration of affected systems and networks.

- **RC.RP-1** – Recovery plan is executed during or after an event – The incident recovery plan is executed during or after an event. During the event, contingencies and workarounds are put in place to restore service levels in the face of a disruption, compromise, or outage. After incident eradication, affected systems should be restored to prior function to fully recovery to a known working state.
- **RC.IM-1** – Recovery plans incorporate lessons learned – Recovery plans should also be updated to incorporate lessons learned.

- **RC.IM-2** – Recovery strategies are updated – Recovery strategies should be reviewed and updated should any improvements be borne out of the analysis of the incident recovery.
- **RC.CO-1** – Public relations are managed – Communications to customers and to the public in general are carefully managed to convey information regarding the incident, status of response and recovery, and planned actions.
- **RC.CO-2** – Reputation after an event is repaired – Typically, the provision of meaningful information regarding the incident and what has been done to recover from the incident helps with preserving reputation but other steps may be necessary.
- **RC.CO-3** – Recovery activities are communicated to internal stakeholders and executive and management teams – Communications to internal stakeholders including executives and management are open and direct regarding the incident, status of response and recovery, and planned actions including evaluation of alternative approaches if the attack persists.

13

DNS APPLICATIONS TO IMPROVE NETWORK SECURITY

Thus far in this book, we've discussed several DNS vulnerabilities and security measures you can take to secure your DNS and network infrastructure, integrity, and information. Securing your DNS improves overall availability of this vital network service required for the basic operation of your IP networks. And the good news is that once you've adequately secured your DNS, you can leverage a secure DNS infrastructure to better secure your overall network and users. In this chapter, we'll discuss some of the ways DNS can be used to help improve your overall security.

DNS inherently lends itself well to "translating" a given piece of information into another related piece of information. This resolution process is the very reason for DNS' invention, and it has been extended beyond resolving domain names into IP addresses and vice versa to support a broad variety of applications. Virtually any service or application that requires translation of one form of information into another can leverage DNS. Understanding the types of resource information stored in DNS and how they are intended for use enables one to consider how they might be exploited and corresponding defense strategies.

Recall that each resource record configured in DNS enables this lookup function, returning a resolution answer for a given query. The DNS server parses the query from the Question section of the DNS message, seeking a match within the corresponding domain's zone file for the query's QNAME, QCLASS, and QTYPE. Each resource record has a Name (aka Owner) field, Class (Internet class is assumed if not specified)

DNS Security Management, First Edition. Michael Dooley and Timothy Rooney.
© 2017 by The Institute of Electrical and Electronic Engineers, Inc. Published 2017 by John Wiley & Sons, Inc.

and Type field. The RData field contains the corresponding answer to the query. The resource record type defines the type and format of the question (owner/name field) and corresponding answer (RData field). Most but not all new applications require new resource record types to enable definition of application-specific information, and these new resource record types are standardized via the IETF RFC process.

Numerous applications are enhanced with DNS, and in this chapter, we'll discuss security applications where DNS plays a key role.

SAFER WEB BROWSING

DNS can facilitate safer web browsing by enabling website publishers to post information about their Transaction Layer Security (TLS) credentials, used to authenticate and encrypt "secure HTTP" traffic. The DNS-based Authentication of Named Entities (DANE) protocol (80, 81) enables access to a website publisher's certificate or certificate authority (CA) information to protect against spoofed certificates or CAs, which can lead to website hijacking unbeknownst to the user/browser.

DNS-Based Authentication of Named Entities (DANE)

Before we dive into DANE, let's review how TLS works to illustrate the vulnerabilities that DANE mitigates. TLS (formerly secure sockets layer, SSL) enables the authentication and encryption of IP traffic as well as data integrity verification. TLS leverages a public key infrastructure (PKI) trust model. Consider Figure 13.1.

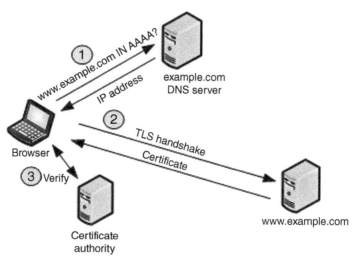

Figure 13.1.   TLS Handshake

The client identifies the web server address using DNS in step 1 as labeled in the figure. In step 2, the client initiates the connection to the server and begins the TLS handshake indicating its supported encryption capabilities and a random number.

The web server uses a private key to sign the random number, passes along the corresponding public key, and presents an ISO X.509 certificate which conveys the server domain name, issuing authority, public key, and other information. If the client (browser) successfully decrypts the random number and is configured to trust the presented certificate, the client may then deem the server trustworthy. This is typically the case within a private network where a trusted certificate is pre-installed on the validating device. In the general case, the client needs to follow the certificate to a trusted CA that has signed the servers' certificate. If the CA verifies the certificate per step 3 in Figure 13.1, the browser trusts the web server and initiates the HTTPS connection.

The link to a trusted root CA, the CA chain of trust, may span multiple layers from the original server to an intermediate authority to a trusted root CA. Each operating system and browser vendor provides a set of several trusted CAs by default. Should the client successfully validate the certificate, it may then commence the secure web session using the server's public key for ensuing communications.

A major vulnerability of this PKI system arises due to the acceptance of a certificate as valid as long as it is confirmed by a configured (trusted) root CA. Since a CA can sign certificates for any child domain, a compromised trusted CA could sign arbitrary certificates for valid server domain names which can lead to false trust in a validated certificate. This can lead to a man-in-the-middle attack where a browser connects to an imposter website with which the user may willingly supply personal or financial data. CA compromise has occurred on a few occasions, such as the Comodo (82), DigiNotar (83), and Symantec (84) attacks and others.

Unfortunately, website administrators have no control over the integrity of CAs or of the list of trusted root CAs installed in browsers. DANE enables website administrators to protect the integrity of their certificate authentication using DNS, and DNSSEC in particular. DANE introduces the TLSA resource record type, which enables a domain administrator to publish in DNS the association of a certificate or public key information with an end entity, for example, the domain name, or trusted issuing authority for the connection endpoint, for example, webserver.

Referring to Figure 13.2, a browser connecting to a website, after having obtained its IP address from DNS (not shown in the figure as step 1) and a certificate from the associated webserver (step 2), verifies the certificate against the compromised CA (step 3), can query for a TLSA resource record for the domain name of the webserver in step 4. The TLSA record enables the DNS administrator to corroborate their webservers' certificate to enable direct validation of the certificate, its public key, or associated root CA. DANE requires DNSSEC validation to assure authentication and data integrity verification.

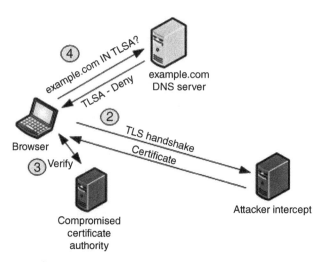

Figure 13.2. Certificate Verification Using DANE

EMAIL SECURITY

DANE can be used to enable domain administrators to publish web addresses and corresponding certificate associations. While originally targeted to secure web browsing,
DANE technology has found wider initial deployment success in securing TLS for
email. An email client may authenticate an email server and communications between
the client and server can be encrypted.

This process works analogously to that just described for HTTPS. An email client
may query for the mail domain's TLSA record for information about the domain's
public keys and certificates, which it can leverage to establish a TLS connection. The
DANE working group of the IETF is still active (85) and is continuing work on defining the association of an S/MIME (Secure/Multipurpose Internal Mail Extension)
user certificate with a domain via a proposed SMIMEA resource record type (86).

Some might view spam email or unsolicited bulk email as a form of attack, and
it has been a nuisance since the dawn of the Internet. Nevertheless, with the unabated
growth of the Internet, the volume of spam emails has seemingly grown even faster.
A variety of techniques exist to combat spam, many of which involve the use of DNS.

Beyond anti-spam initiatives, the very security of your users' email transactions
is likewise at risk of interception, redirection, and corruption where attackers can
disrupt communications and gather sensitive information. Several attack vectors are
initiated via email to induce users to click a link or attachment to initiate malware
installation, for example.

The National Cybersecurity Center or Excellence (NCCoE) recently published
an excellent draft practice guide (87) which provides use cases, configuration
examples, and testing outcomes for a variety of email transaction scenarios that

leverage much of the technology discussed in this section. Before we delve into this technology, let's first look at the anatomy of an email transmission including the role of DNS in email delivery.

Email and DNS

An email typically originates from one person and is sent to one or more recipients. Each email address is formatted as mailbox@maildomain. The mailbox commonly refers to the name of the person or owner of a mailbox or email account, while the maildomain, typically the company or Internet provider name, is the destination domain for delivery to the corresponding mailbox or mail exchange. Emails are composed using an email client, such as Microsoft Outlook, Eudora, or web-based clients like Yahoo and Google. Regardless, when sent by the originator, the client connects to a Simple Mail Transfer Protocol (SMTP) server (using the SMTP protocol) to send the email. Like a default router for email, the SMTP server is responsible for forwarding the email to its destination.

The SMTP server must resolve the maildomain to an IP address for transmission of the message. Naturally this is done using DNS with a lookup for the Mail Exchange (MX) record type, as well as the corresponding A or AAAA record types.

MX – Mail Exchange Record   The mail exchange record is used to locate an email server or servers for a particular domain. If I send an email destined to tim@ipamworldwide.com, my SMTP server will use DNS to find the host(s) that can receive emails for users in the ipamworldwide.com domain. More than one MX record may be created per domain, and each can be defined with a different preference value. Use of the preference field enables the sending SMTP server to prioritize the destination host to which it will forward the email for the given domain, and if unavailable to a second (and third, etc.) choice destination. The lower the preference value, the more preferred the listed destination. In the example below, we have two MX records for the ipamworldwide.com domain. The destination smtp1 is preferred (lower preference) over smtp2. However, if smtp1 is unavailable, this mechanism provides a backup server for email delivery.

Owner	TTL	Class	Type	RData	
Email destination domain	TTL	IN	MX	Preference	Mail server host domain name
ipamworldwide.com.	86400	IN	MX	10	smtp1.ipamworldwide.com.
ipamworldwide.com.	86400	IN	MX	20	smtp2.ipamworldwide.com.

Note that the mail server host domain name within the RData field must have a corresponding A or AAAA record to complete the required resolution to a reachable

IP address. Many DNS servers supply these address records within the Additional section of the MX query response.

Upon resolving the destination mail server, the SMTP server sends the message to the destination using the SMTP protocol. The ultimate destination server, to which recipient email clients connect, must support Post Office Protocol (POP) or Internet Message Access Protocol (IMAP) to enable client retrieval of the email message. Thus when your email client performs a "send/receive," it utilizes SMTP to send outgoing messages to its configured SMTP server and POP or IMAP to retrieve incoming email messages from the configured POP/IMAP server.

Figure 13.3 highlights a very simple SMTP transaction between two email servers, shown on opposite sides of the Internet cloud. On the left of the figure, Mike composes an email to tim@ipamworldwide.com using his email client and sends

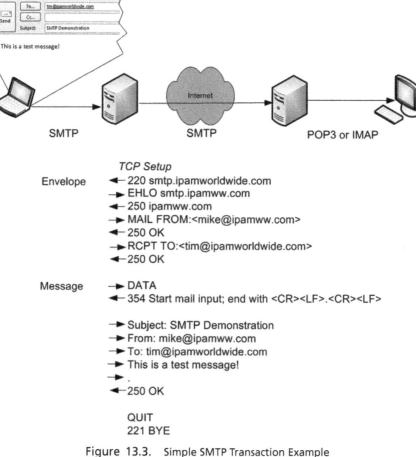

Figure 13.3. Simple SMTP Transaction Example

it. His configured SMTP server forwards the message to the destination server, as resolved by the MX record(s) for ipamworldwide.com. His SMTP server initiates a TCP connection on port 25 with the resolved destination server. The message transactions between the two servers are highlighted in the figure.

Once the TCP session is established, the SMTP application utilizes the session to handshake and process the message. The envelope portion of the message begins with the HELO (or EHLO, enhanced HELO) which conveys the sending entity's identity. The MAIL FROM statement indicates the source of the message, followed by the RCPT TO statement indicating the destination mailbox. At this point in the exchange, the recipient server may refuse to accept the message and close the connection if the destination mailbox is unknown or blocked, or if the "from address" is prohibited. Otherwise, the transaction continues and the data or message portion* is transmitted. The receiving mail exchange stores the email message or forwards it to the server on which the destination mailbox resides.

The store-and-forward approach used by the received email server may also be used by intermediate email gateways (aka message transfer agents) to provide multi-hopped email delivery. As mentioned above, the resolution of a destination mailbox domain to multiple MX records implies this ability to identify a "destination" mail server which may or may not be the final destination from which the intended recipient retrieves the email. The MX record preference field provides control over the relative preference of incoming mail servers or gateways, while providing selection from among multiple choices based on availability and performance.

Figure 13.4 illustrates a two-step email delivery scenario using SMTP. In this scenario, the same email is being sent as shown in Figure 13.3. However, in this case, perhaps the intended destination server, smtp.ipamworldwide.com is busy and refuses a direct connection. Having resolved both the ipamworldwide.com server and an mta-gateway.com server via a DNS MX query, my outgoing mail server will attempt to send the email to the second choice, mta-gateway.com.

In accepting the SMTP transmission from my mail server, the mta-gateway.com server effectively agrees to forward the email to the ultimate destination on my behalf. The transaction between my mail server and the mta-gateway.com server completes before the second leg of transmission is attempted. SMTP uses a store-and-forward approach, not synchronous relaying of each message.

The first leg of the transmission looks very similar to that of Figure 13.3, except for the difference in the SMTP server. The second leg of the connection is also similar, except once again for the SMTP endpoints. The other difference is the insertion of the *Received:* line within the header portion of the data section of the mail. Each intermediate SMTP server which forward the message prefixes a "Received" line indicating its domain name and corresponding timestamp. This enables tracing

---

* Note that the message portion of an email consists of a header and the body. As a point of reference, RFC 5321 (125) defines the SMTP specification, while RFC 5322 (126) defines the Internet message format for email, defining valid header and data syntax.

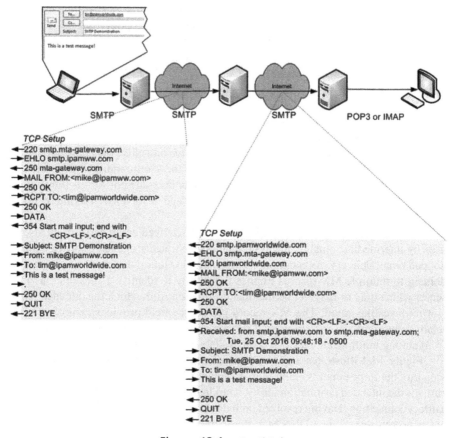

Figure 13.4.   Email Relay

of the email from the destination back to its path. The RCPT TO line remains the same in both segments, indicating the mailbox to which errors in delivery should be sent.

As footnoted above, the message portion of an email consists of a header and the body. Each header field consists of a word followed by a colon and a value. The header contains a variety of data including the following:

- Originator fields: from, sender, reply-to, orig-date
- Destination fields: to, cc, bcc
- Identification fields: message-id, in-reply-to, references, msg-id, id-left, id-right, no-fold-quote, no-fold-literal
- Informational fields: subject, comments, keywords

- Resent fields (informational fields relating to the reintroduction* of a message into the Internet, e.g., by a emailing service): resent-date, resent-from, resent-sender, resent-to, resent-cc, resent-bcc, resent-msg-id
- Source trace information: trace, return, path received, name-val-list, name-val-pair, item-name, item-value

We have summarized the basic email process and types of information that may be included in a given email message because different anti-spam techniques utilize different information sources in validating the sender as a legitimate or acceptable sender of emails. We'll discuss those techniques that utilize DNS to perform this validation next.

DNS Block Listing

The use of white or black listing (88) provides a simple means for the recipient email server to lookup a sender's IP address via DNS and to validate its legitimacy. Block list providers track IP addresses known to originate spam email and publish this information in DNS to facilitate a simple DNS lookup upon email receipt to determine if the email should progress to its recipient or be discarded. This lookup is typically formed by reversing the IP address of the source IP address of the email message, just as is done in forming PTR records. Note that the source IP address being analyzed is that from which the email was received directly, perhaps an email gateway, which may or may not be the original transmitter. However, the intent of such listing is to identify such senders of email by IP address as legitimate or not.

In this scenario, the reversed IP address is appended with a given domain name, typically that of the black list provider. The "host domain name" thus formed by this concatenation comprises the Qname which is queried in DNS using the A resource record query type, not PTR. The query answer is interpreted based on whether the record was found, in which case often an IP address within the 127/8 block is returned, and on whether the list publishes known spammers (black or block list) or known non-spammers (white list).

For example, upon receiving an email message with a source IP address of 192.0.2.95, my email server formulates an A record query for hostname 95.2.0.192.spamblocklist.org, assuming my chosen black list provider publishes lookups within the spamblocklist.org domain. Upon receiving a reply with answer (IP address) 127.0.0.5, the email server classifies the email as spam and rejects it. On the other hand, if NXDOMAIN is returned for the query, the email may be permitted. A whitelist service, publishing known genuine email server addresses would render the opposite interpretation based on the DNS lookup.

---

* Reintroduction is not forwarding. The transmission of an email with the *original sender* information instead of that of the transmitter is considered reintroduction. Forwarding uses the mailbox doing the forwarding as the sender.

Sender Policy Framework (SPF)

The Sender Policy Framework (SPF) is currently defined in RFC 7208 (89). SPF enables an organization to publish its own list of authorized outgoing email server addresses, a self-published white list, though with substantially more sophistication. Under SPF, the received email message's envelope information is examined, and a TXT DNS query from the email recipient is based upon the sender, the sender's domain, as well as the sender's source IP address. The TXT record is encoded as a string of "mechanisms" that are used to process the source IP address from which the email originated, the domain portion of the MAIL FROM or HELO identity, and the sender parameter from the MAIL FROM or HELO identity.

The SPF attempts to provide validation of what hosts are configured to send email for a given domain. That is, SPF seeks to eliminate spam emails from spoofed domains purporting to be from the SPF publisher's domain. A recipient email host can look up the TXT record for the sender's domain to verify that the sending email host matches those authorized by the sender.*

As mentioned, SPF utilizes the TXT RRType with a particular syntax for SPF interpretation per RFC 7208. The syntax includes a version string (v=spf1) followed by a space, then one or more terms that define qualifiers on resource record types or IP network addresses, modifiers, and even macros. We enumerate these policy parameters in detail next.

Owner	TTL	Class	Type	RData
Domain name	TTL	IN	TXT	Version, directives and/or modifiers
smtp.ipamworldwide.com.	86400	IN	TXT	v=spf1 +ip4:192.0.2.32/30 –all

Mechanisms   Mechanisms enable specification of the match criteria within the TXT record which a receiving email server can query to validate the sender of a given email message. Mechanisms are defined within the TXT record's RData field after specification of the SPF version. Mechanisms are evaluated left to right. If a mechanism passes based on evaluation of the mechanism, the verification passes; otherwise, the next mechanism is tested until a pass or fail is found or no further mechanisms are defined.

Each mechanism can be defined with a qualifier, a prefix which instructs the mail or spam filter server how to interpret a given "match."

- + = pass (default) – consider this mechanism a pass if this mechanism matches
- − = fail – consider this mechanism a fail if the mechanism matches

---

* Sender ID is a related spam detection technique that also uses the TXT resource record type though it analyzes different information from an incoming email message and has been less popular in its implementation (127) so we won't delve into Sender ID here. Please consult (8) for details on Sender ID.

- ~ = soft-fail – consider this mechanism somewhere between neutral and fail if this mechanism matches; this interpretation would not fail this check outright if it matched, but would hold it for closer scrutiny
- ? = neutral – consider this mechanism neutral if this mechanism matches

Qualifiers may be used with the following resource record check based mechanisms to define the interpretation of a given mechanism as shown in the examples following:

- a = lookup the A record for the source domain (from the MAIL FROM or HELO identity); if it matches the source IP address of the message, this mechanism matches. This can be scoped to a specific domain and/or number of CIDR bits to compare in the addresses as illustrated in the following examples:
  - +a = pass if the A record query for the source domain matches the source IP address
  - –a:ipamworldwide.com = fail if an A record query for ipamworldwide.com matches the source IP address
  - ~a/24 – soft-fail if the first 24 bits of the IP address retrieved via A record lookup of the source domain matches the first 24 bits of the source IP address
- mx = lookup the MX record for the source domain (from the MAIL FROM or HELO identity); for each MX lookup resolved, look up the corresponding A record; if it matches the source IP address of the message, this mechanism passes. As with the A record mechanism, the mx mechanism can be scoped to a specific domain and/or number of CIDR bits to compare in the addresses as illustrated in the following example:
  - +mx:ipamworldwide.com/28 = pass if an A record associated with an MX record lookup is returned where the first 28 bits match the first 28 bits of the source IP address of the message
- ptr = (note that this mechanism, while still included in the specification, is not recommended for use); lookup the PTR record (up to 10) corresponding to the source IP address of the email message; then compare two things with each domain name returned in the PTR lookup
  - Check that the domain name returned matches the source domain of the email message
  - Check that the corresponding A or AAAA record returns an IP address matching the source IP address

    If both conditions hold, this mechanism passes. This mechanism can be further scoped by a domain name, which can be used to filter multiple returned PTR-lookup domain names as illustrated in the following examples:
  - –ptr – fail if a domain name returned during the PTR lookup of the source IP address matches the source domain and if the A/AAAA domain name

corresponding to the domain name returned during the PTR lookup matches the source IP address of the email

- o +ptr:ipamworldwide.com – pass if a domain name returned during the PTR lookup of the source IP address matches the source domain while falling within the ipamworldwide.com domain and if the A/AAAA domain name corresponding to the domain name returned during the PTR lookup matches the source IP address of the email
- ip4 = verify that the source IP address matches the IPv4 address specified; this mechanism may be qualified by CIDR length as illustrated in the following example:
  - o ?ip4:192.0.2.32/30 – neutral if the source IP address of the message falls within 192.0.2.32-192.0.2.35
- ip6 = verify that the source IP address matches the IPv6 address specified; this mechanism may be qualified by prefix length as illustrated in the following example:
  - o +ip6:2001:db8:f02b:2a::/64 – pass if the source IP address of the message falls within the 2001:db8:f02b:2a::/64 network
- exists:*domain_name* = lookup the A record (not AAAA) corresponding to the *domain_name*; this mechanism matches if any answer (IP address) is provided (this mechanism must be scoped with a domain name to match as illustrated in the following example)
  - o exists:ipamworldwide.com – matches if an A record lookup for the ipamworldwide.com domain returns an IP address
- include:*domain_name* = recursively evaluate the *domain_name* to leverage its SPF policies, for example, to utilize the policy of a domain from multiple ISPs or from other domains from which you send email
- all = matches everything; often used as the final parameter as –all to fail if no prior mechanism matches

Modifiers   Modifiers may be specified within SPF records to provide additional information. Modifiers are name-value pairs, two of which have yet been defined.

- redirect=*domain_name* – enables "aliasing" of SPF records, for example, to apply a common SPF processing record to multiple domains. This provides a convenience for ongoing change management: change the processing in one record, minimizing errors and maximizing consistency. In the following example, the MX record check for the ipamworldwide.com domain would apply to the hq and euro subdomains as well.

```
hq.ipamworldwide.com.    IN SPF "v=spf1 redirect=_spf.ipamworldwide.com"
euro.ipamworldwide.com.  IN SPF "v=spf1 redirect=_spf.ipamworldwide.com"
_spf.ipamworldwide.com.  IN SPF "v=spf1 +mx:ipamworldwide.com -all"
```

The redirect can be used explicitly as in the above example, or as a "last resort," for example, listed as the rightmost mechanism.

- exp=*domain_name* – explanation, which defines the domain for which a TXT record lookup must be done to identify the string to be presented as results upon a mechanism match failure.

Macros   Technically, the *domain_name* for any of the above mechanisms and modifiers need not be an explicitly defined (hard coded) domain, but one that can be defined using macros to dynamically formulate a domain name based on the message envelope under evaluation. Even the TXT record fetched by processing an exp modifier may be populated with macros. Macros are identified using the percent sign (%) and are enclosed in curly brackets({}). The following macros have been defined:

- s = the sender's email address
- l = the local part of the sender's email address
- o = the domain of the sender's email address
- d = the current domain, usually the same as the sender's domain but may also have been processed, for example, via the include mechanism
- i = the source IP address of the message sender
- p = the validated domain name via PTR lookup of the source IP address of the message sender
- v = the literal string "in-addr" if the source IP address is an IPv4 address and "ip6" if the source IP address is IPv6.
- h = the domain part of the HELO/EHLO identity
- %% = the literal %
- %_ = space " "
- %- = a URL-encoded space, for example, "%20"

The following macros are available for use in the TXT record referenced by an exp mechanism and may not be used elsewhere:

- c = the SMTP client IP address
- r = the domain name of the host performing the SPF check
- t = the current timestamp

Macro transformers enable use of a subset of the results of a macro, for example, by specifying an integer quantity of domain name labels, or the reversal of the results of a macro, for example, reversing an IP address. Reversal is performed by adding an r into the macro curly brackets.

Macro Examples Consider the example of Figure 10-3, where Mike (mike@ipamww.com) sends an email to tim@ipamworldwide.com from the SMTP host on IP addresses 192.0.2.32. Using this and other information from the figure, we can define the macro values for this email transmission as

- $\%\{s\}$ = mike@ipamww.com
- $\%\{l\}$ = mike
- $\%\{o\}$ = ipamww.com
- $\%\{d\}$ = ipamww.com
- $\%\{d3\}$ = ipamww.com
- $\%\{d2\}$ = ipamww.com
- $\%\{d1\}$ = com
- $\%\{i\}$ = 192.0.2.32
- $\%\{ir\}$ = 32.2.0.192
- $\%\{v\}$ = in-addr
- $\%\{h\}$ = ipamww.com
- $\%\{ir\}.\%\{v\}._spf.\%\{d\}$ = 32.2.0.192.in-addr._spf.ipamww.com

SPF provides a powerful macro language to granularly articulate email policies for your organization.

Domain Keys Identified Mail (DKIM)

Domain keys identified mail (DKIM) specifies a means for a sender of email to cryptographically sign an email message such that recipients may validate it upon receipt via retrieval and application of the sender's domain key. DKIM supports data origin authentication and data integrity verification through the use of digital signatures. This enables the originator of a given set of data (an email message in this case) to sign the data such that those receiving the data and the signature, along with a corresponding public key can decipher the signature. Like DNSSEC, DKIM employs an asymmetric key pair (private key/public key) model to provide email origin authentication and data integrity verification without data privacy through encryption.

Recalling our discussion of such a model in Chapter 8, the signed information (email message and selected header fields) are encrypted with a private key by the sender and can be validated by the recipient by decrypting the data with the corresponding public key. This provides authentication that the data verified was indeed signed by the holder of the private key. Digital signatures also enable verification that the data received matches the data published and was not tampered with in transit.

The email originator signs the message with its private key and the message and its associated signature are transmitted to the recipient. A new email header, DKIM-Signature, has been defined to store the DKIM signature with information on retrieving the public key. Based on our prior review of how SMTP works, you may be wondering how modification of envelope data and insertion of headers affect the signature. DKIM offers a "simple" or strict form of canonicalization and a "relaxed" form. The simple form tolerates very little modification while the relaxed form permits white space replacement and header line rewrapping without impacting the signature validity.

DKIM Signature Email Header Field   The recipient must extract the signature from the `dkim-signature` header field. The DKIM-Signature field also contains

- The DKIM version (e.g., `v=1`)
- The algorithm used to generate the signature (e.g., `a=rsa-sha256`)
- Signature (e.g., `b=dqdVx0fAK9…`)
- Hash of the canonicalized message body (`bh=7Dkw0eE35Jlkjexcmpol…`)
- Canonicalization method (`c=relaxed`)
- The signing domain identifier – the domain of the signing entity (e.g., `d=ipamworldwide.com`)
- User or agent on whose behalf the message is signed (`i=tim @ipamworldwide.com`)
- The selector or key reference within the domain (allows multiple keys per domain which aids in key rollover and more granular signatures) (e.g., `s=europe`)
- Enumeration of the header fields that were signed (e.g., `h=from:to: subject:date`)
- Additional optional information, including query methods to use to retrieve the public key. The default (and currently only) query method, `q=dns/txt`, instructs the recipient to perform a DNS query of querytype "txt" to retrieve the public key that corresponds with the private key that was used to sign the message. Another optional field of interest, the `i=` tag provides the identity of the user or agent on whose behalf this message was signed.

DKIM TXT Record   Using the header-specified query method q=dns/txt, the recipient performs a DNS query for a TXT record for the signing domain. The question section of the query is formulated by concatenating the selector value (s= value), the string "_domainkey" and the specified signing domain (d= value). Using the example where s=europe and d=ipamworldwide.com as specified in the dkim-signature field of an incoming email, a TXT query for `europe._domainkey.ipamworldwide.com` would be issued. The RData portion of the corresponding TXT record includes one or more tags similar to the DKIM-Signature field.

- DKIM version (`v=DKIM1`)
- Granularity of the key, which if specified, must match the local part of the identify flag in the DKIM-Signature header (`g=*`)
- Hash algorithm(s) accepted (e.g., `h=sha256`)
- Key type (`k=rsa`)
- Notes for human consumption (`n=updated_key`)
- The public key (`p=Dkjeijf8d98Kz…`)
- Service type (`s=email`)
- Flags indicating such things as the compliance rules among the `i=` tag in the DKIM-Signature header and the `d=` domain tag (encoded in the TXT record as `t=s`), as well as whether this domain is testing DKIM (`t=y`).

The only required tag is the p tag, the public key. An example TXT record follows:

```
europe._domainkey.ipamworldwide.com  IN TXT ("v=DKIM1; p=Dkjeijf98Kz…")
```

Upon retrieving the public key, the recipient computes a hash of the received message body and signed header fields, as did the originator. The recipient applies the hash algorithm to the received signature using the originator's public key. The output of this decryption, the original data hash, is compared with the recipient's computed hash of the data. If they match, the data has not been modified and the private key holder signed the data.

If an incoming email message contains a DKIM-Signature header field, it's clear that the sender is using DKIM and has signed the message. But if an incoming email message does not contain a DKIM-Signature header field, does this mean the sender does not sign messages? This in fact could create an opening for a SPAM attacker issuing unsigned email messages from a spoofed source domain. DKIM relies on publication of Author Domain Signing Practices (ADSP), which enables a recipient email server to determine whether the message from a given domain by policy should be signed and if so, by whom and with what signature(s).

A recipient determines the sending domain's signing practices by issuing a query for Qtype=TXT and Qname = `_adsp._domainkey.`*`signing-domain-identifier`*, where *`signing-domain-identifier`* is again the d= value. The corresponding TXT record indicates whether email from this domain is always signed, may be signed, and is always signed and any unsigned email should be discarded. Please refer to RFC 5617 (90) for additional details.

Domain-Based Message Authentication, Reporting, and Conformance (DMARC)

Domain-based message authentication, reporting, and conformance (DMARC) builds upon and works in conjunction with SPF and DKIM. DMARC seeks to improve information exchange between an email sender and receiver beyond that provided by SPF and DKIM in order to improve validation, message disposition, and sender feedback. Documented in RFC 7489 (91), DMARC enables domain owners to publish email policy assertions about their domains via DNS, while enabling email receivers to authenticate senders, determine email disposition, and report feedback to the sender.

DMARC policies are published in the DNS in the form of TXT records. Upon receipt of emails from the domain, the receive can query for TXT records for the sender's domain. The format of the TXT record for DMARC utilizes the following concatenation of policy tags defined as follows.

- v – protocol version, for example, v=DMARC1
- pct – percentage of messages subjected to filtering – value between 0 and 100
- ruf – reporting URI for forensic reports
- rua – reporting URI for aggregate reports
- p – policy for organizational domain – none, quarantine, or reject
- sp – policy for subdomains of the organizational domain
- adkim – alignment mode for DKIM – relaxed or strict
- aspf – alignment mode for SPF – relaxed or strict
- fo – failure reporting options – generate a failure report under one of the following conditions:
  o DMARC failure report if all underlying authentication mechanism fail to produce a "pass" result
  o DMARC failure report if any underlying authentication mechanism fail to produce a "pass" result
  o DKIM failure report if the message had a signature that failed to validate
  o SPF failure report if the message failed SPF evaluation
- rf – reporting format for use when both SPF and DKIM fail in order to provide details for each
- ri – reporting interval in seconds

Figure 13.5 illustrates how these policies are applied. The sender on the left creates an email and sends it. The sender email system inserts a DKIM header and transmits the email via SMTP to the recipient email system. The recipient system may perform standard block listing, rate limiting, etc. as a first line of defense. Next the sender policy TXT records are retrieved for the sending domain which should

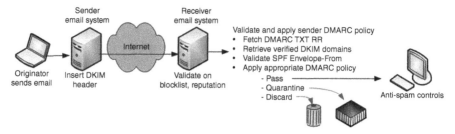

Figure 13.5. DMARC Email Policies

indicate SPF and DMARC policies. The receiving email system then validates the DKIM sending domain, the SPF on the envelope, then applies the DMARC policy based on the results of these prior validation steps.

If validation passes, the email is passed on to the recipient, who may apply additional anti-spam, blocking policies. If the DKIM and/or SPF validation fails, the DMARC policy shall dictate the desired action, whether to quarantine or discard the email. In this failed validation scenario, the receiving email system should store this disposition for future reporting to the sender according to the DMARC reporting policy. This process reduces ambiguity in handling emails failing validation and enables a feedback loop to the sender regarding the malformed or potential imposter email.

SECURING AUTOMATED INFORMATION EXCHANGES

Beyond signing updates, DNS provides an additional mechanism to validate the originator of a DNS update message. This is helpful in preventing an attacker from modifying your "www" record through a dynamic DNS message, for example. Of course this example is a bit extreme given your external authoritative DNS servers should never allow DDNS updates, but consider the implications of a change, malicious or erroneous, to a critical server in your infrastructure.

Dynamic DNS Update Uniqueness Validation

DHCID – Dynamic Host Configuration Identifier Record   Dynamic DNS enables the updating of DNS information with DHCP clients' assigned IP address information. Thus a DHCP server on behalf of the client or the client itself can update DNS with the client's IP address and hostname association via A/AAAA and PTR* records. It is quite possible that the same hostname/FQDN may be claimed

---

* Associating client identification information with PTR records is not currently specified in the DHCID RFC.

by multiple DHCP clients, or that a client may claim a hostname already assigned to a predefined (e.g., statically addressed) device.

The DHCID record provides client identification information in DNS to uniquely associate the particular DHCP client with the hostname/FQDN being updated by the DHCP server. The DHCID record would be defined in the Pre-requisite section of the DNS Update message to verify the record "owner" for updating.

The DHCID record uses the same owner field as the corresponding A or AAAA record. The RData portion of the record is formed by performing a one-way secure hash using the SHA-256 algorithm over the following concatenated fields:

- Identifier Type code (2 bytes) – identifies the information within the DHCP packet that was used in creating this hash. Possibilities include client hardware address, client identifier option, or device unique identifier (DUID).
- Digest Type code (1 byte) – identifies the hash algorithm. The RFC defines values of 0 (reserved) or 1 (SHA-256) though IANA maintains a registry for future value assignments.
- Digest of the data from the DHCP packet as identified by the Identifier value concatenated by the client's FQDN.

Owner	TTL	Class	Type	RData		
				Identifier	Digest	SHA-256 hash of
Host domain name	TTL	IN	DHCID	Type	Type	{identifier type, fqdn}
w3.ipamworldwide.com.	86400	IN	DHCID	A1B87Y2/AuCcg8e93aQcjl...		

STORING SECURITY-RELATED INFORMATION

As a scalable distributed repository that is universally accessible, at least to the extent you permit such access, DNS can store data other than IP address and domain name information; even security oriented information.

Other Security Oriented DNS Resource Record Types

TA – Trust Authority Record   While an RFC does not exist defining the TA resource record, IANA has assigned it a value, so we'll mention it here. The TA resource record is identical in format to the DS record type including RData fields for key tag, algorithm, digest type, and digest. Use of the TA record enables a resolver to have a resource record signature validated by a known trust authority even if the root zone had not been signed. This functionality was superseded by the DLV record but with the root zone being signed both the TA and DLV records have little utility on the Internet at large.

CERT – Certificate Record RFC 4398 (92) defines the CERT record as a means to store certificates and certificate revocation lists (CRLs) in DNS. Certificates provide a means to identify an organization, server, individual, or other entity and associate a public key with that identity. The public key can be used to authenticate the sender's identity and to encrypt and decrypt communications and validate message integrity. Certificates are hierarchical and can be used to validate up to a known trusted entity (certificate authority). CRLs are lists of certificates which have been revoked due to expiration or manual revocation.

CERT records containing certificates are stored in DNS to enable resolvers to obtain certificates via DNS instead of from a destination certificate server. The CERT resource record has the following format.

Owner	TTL	Class	Type	RData			
Domain name	TTL	IN	CERT	Certificate Type	Key Tag	Algorithm	Certificate or CRL
ipamww.com.	86400	IN	CERT	PGP	436	3	A4df480DFC9lLa....

The owner field identifies the entity to which the certificate applies when a certificate is included in the RData portion of the record. If a CRL is included in the RData section, the owner name should contain the domain name related to the issuing authority. The RData portion contains the following subfields:

- Certificate Type such as X.509/PKIX, PGP, and others
- Key Tag, which is used to streamline the identification of relevant certificates to those of matching key tags
- Algorithm – the algorithm used in generating the key, which is encoded in the same manner as the Algorithm field of the DNSKEY resource record type
- The certificate or CRL

As we discussed at the beginning of this chapter, DANE provides the ability to verify certificates via the TLSA resource record, which also provides further flexibility in defining associated certificate information. DANE also requires DNSSEC which authenticates the integrity of the TLSA answer as well as origin authentication.

IPSECKEY – Public Key for IPSec Record The IPSECKEY resource record type, defined in RFC 4025 (93), provides a means to store a public key in DNS for use with IPSEC. This resource record enables a client seeking to establish an IPSec tunnel to a remote host to identify a means to authenticate the remote host and to determine whether to connect directly to the host or connect via another node acting as a gateway. IPSECKEY resource records are associated with the intended remote

host's IP address or host domain name. IP addresses are stored in the .arpa. reverse
domain space. The format of the IPSECKEY resource record is as follows:

Owner	TTL	Class	Type	RData					
IP address in .arpa. domain or host domain name	TTL	IN	IPSECKEY	Prece- dence	Gateway Type	Algo- rithm	Gateway	Public Key	
1.0.12.10.in-addr.arpa.	86400	IN	IPSECKEY	10	1	2	10.100.1.2	Adf4C9lL....	

The RData field contains the following fields:

- Precedence – used to prioritize multiple records within a common RRSet, using
  the lowest precedence first.
- Gateway Type – indicates the format of the Gateway field
  - 0 = no gateway is present
  - 1 = IPv4 address
  - 2 = IPv6 address
  - 3 = FQDN
- Algorithm – the format of the Public Key field
  - 0 = no key is present
  - 1 = DSA formatted key
  - 2 = RSA formatted key
- Gateway – identifies a gateway to which an IPSec tunnel can be established to
  reach the remote host (identified by the owner field). The interpretation of this
  field is governed by the Gateway Type field
- Public Key – the key generated using the algorithm specified in the Algorithm
  field

KEY – Key Record   The KEY record was defined with the initial incarnation
of DNSSEC, but was superseded by the DNSKEY resource record. However, prior
to the release of the current incarnation of DNSSEC, DNSSEC*bis*, the KEY record
was also utilized to store public keys associated with the SIG(0) record. The KEY
record has the same format as the DNSKEY record.

Owner	TTL	Class	Type	RData			
Key name	TTL	IN	KEY	Flags	Proto- col	Algo- rithm	Key
K3941.ipamww.com.	86400	IN	KEY	256	3	1	12S9X-weE8F(le...

KX – Key Exchange Record   The KX record enables specification of an intermediary that can supply a key on behalf of another host. In other words, if intending to perform key negotiation with x.ipamworldwide.com, the KX record could point to the y.ipamworldwide.com host domain name with whom key exchange negotiation should ensue. A preference field enables specification of multiple alternate domains of varying preference for key negotiation.

Owner	TTL	Class	Type		RData
Host domain name	TTL	IN	KX	Preference	Key exchange host domain name
x.ipamworldwide.com.	86400	IN	KX	10	y.ipamworldwide.com.
x.ipamworldwide.com.	86400	IN	KX	20	z.ipamworldwide.com.

SIG – Signature Record   The SIG resource record has been superseded by the RRSIG record within the scope of DNSSEC, though the SIG record is still in use for digitally signing DNS updates and zone transfers outside the scope of DNSSEC. That is, you don't use DNSSEC to enable transaction signatures of updates and zone transfers. Such transactions can be signed using shared secret keys via TSIG (Transaction Signature) records or by using private/public key pairs via SIG(0), where corresponding public keys are stored as KEY records. The notation SIG(0) refers to the use of the SIG resource record with an empty (0) Type Covered field. In such cases, RFC 2931 (94) recommends setting the owner field to root, the TTL to 0, and class to ANY as shown in the example below.

The SIG record is formatted identically to the RRSIG record, with the exception of the formatting of the Expiration Date and Inception Date fields; for the SIG record, these fields are not formatted by date per the RRSIG record and are instead formatted as an incremental integer, enumerated as the number of seconds since January 1, 1970 00:00:00 UTC. This counter will rollover to 0 and continue counting after the counter exceeds 4.29 billion seconds (a little over 136 years).

Owner	TTL	Class	Type				RData					
RRSet Domain	TTL	IN	SIG	Type Cov.	Alg.	Labels	Orig. TTL	Expire	Inception	Key tag	Signer	Signature
.	0	ANY	SIG	0	3	3	86400	2016051 5133509	2001611 5133509	26421	ipamww.com.	Zx9v…

SSHFP – Secure Shell Fingerprint Record   The Secure Shell (SSH) protocol enables secure login from a client to a server and other secure network services over an insecure IP network. The security of the connection relies upon the user authenticating him- or herself to the server as well as the server authenticating itself to the client via Diffie–Hellman key exchange. If the public key is not already known by the client, a fingerprint of the key is provided by the server for verification by the user. Storage of this key fingerprint in DNS provides a means for the client to lookup and verify the fingerprint out of band via a "third party." The lookup requires use of DNSSEC to secure the lookup process and assure message integrity. The SSHFP resource record is the record type used to store these SSH fingerprints.

Owner	TTL	Class	Type	RData		
Host domain name	TTL	IN	SSHFP	Algorithm	Fingerprint Type	Fingerprint
srv21.ipamww.com.	86400	IN	SSHFP	2	1	8Fd7q90D+fd…

The RData portion of the SSHFP record includes the following fields:

- Algorithm – currently defined values are
  - 0 = Reserved
  - 1 = RSA
  - 2 = DSA
- Fingerprint Type – currently defined values are
  - 0 = Reserved
  - 1 = SHA-1
- Key fingerprint

SUMMARY

DNS has proven remarkably effective, scalable, and versatile over the decades as the Internet's directory. Protecting your network through the use of DNS and through the protection of DNS infrastructure and data is critical to effectively managing risks to your organization. Web browsing and email are among the top applications your users require for daily activities. Securing your DNS not only protects your DNS but also enables you to better secure these and other key application transactions.

14

DNS SECURITY EVOLUTION

This book has detailed vulnerabilities and risks for your DNS implementations as well as steps to mitigate your exposure. These risks apply not only to your DNS infrastructure but also to your network and computing infrastructure as a whole. The DNS service is vital to the very operation of your IP network and Internet access. Thus, maintaining an "always-on" DNS service is crucial to the integrity of your network. Bad actors know this and often rely on the availability of DNS to perform illicit activities to disrupt your network or steal sensitive information.

Many organizations take DNS for granted and don't even monitor DNS traffic today. However, you can't detect or mitigate what you can't see. The focus of network security traditionally lies with in-band technologies such as firewalls, network access control, intrusion detection/ prevention systems, and the like. But DNS is a foundational protocol freely operating across your network. We advise you to supplement your network security strategy to incorporate DNS data, analysis, and intelligence if you have not already done so. DNS is used before in-band communications commence and may be instrumental in early attack detection.

Figure 14.1 illustrates our view of a DNS maturity model based on the NIST Cybersecurity Framework tiers. As we discussed in Chapter 1, these tiers reflect an increasingly security savvy organization in general. We've applied DNS-specific attributes to these tiers in Figure 14.1, focusing primarily on the "protect" and "detect" security categories.

DNS Security Management, First Edition. Michael Dooley and Timothy Rooney.
© 2017 by The Institute of Electrical and Electronic Engineers, Inc. Published 2017 by John Wiley & Sons, Inc.

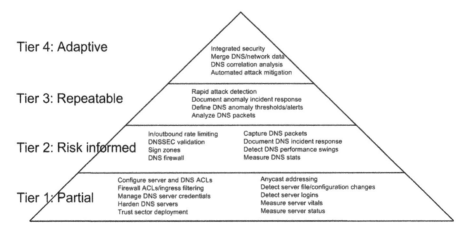

Tier 4: Adaptive

Integrated security
Merge DNS/network data
DNS correlation analysis
Automated attack mitigation

Tier 3: Repeatable

Rapid attack detection
Document anomaly incident response
Define DNS anomaly thresholds/alerts
Analyze DNS packets

Tier 2: Risk informed

In/outbound rate limiting
DNSSEC validation
Sign zones
DNS firewall

Capture DNS packets
Document DNS incident response
Detect DNS performance swings
Measure DNS stats

Tier 1: Partial

Configure server and DNS ACLs
Firewall ACLs/ingress filtering
Manage DNS server credentials
Harden DNS servers
Trust sector deployment

Anycast addressing
Detect server file/configuration changes
Detect server logins
Measure server vitals
Measure server status

Figure 14.1. DNS Security Maturity Model

Tier 1 reflects a partial or ad hoc security implementation. Implementations of the measures defined here reflect the foundation for securing your DNS. The defensive measures included in this tier include the following:

- Trust sector deployment of your DNS infrastructure
- Hardening of server and hardware controls
- Basic authentication of DNS update, notify, and zone transfers using ACLs and Transaction signatures

Anycast addressing is an effective DDoS defensive measure and may also be implemented by organizations in this tier.

Tier 1 security steps are the minimum an organization should implement in order to provide for resilient and performant DNS services and to have the ability to detect outages and potential DoS/DDoS forms of attack. From a detection standpoint, organizations at Tier 1 monitor server fundamentals like daemon process status and hardware metrics such as CPU, memory, disk and I/O utilization. Most organizations today have implemented at least some or all Tier 1 security measures but may lack full documentation of threat detection, response and recovery procedures, and organization-wide awareness.

The second tier builds on Tier 1 with measures to protect against attacks more sophisticated than DoS/DDoS attacks. Implementation of DNSSEC, both in configuring DNSSEC validation for recursive servers and DNSSEC zone signing for authoritative servers can help improve the integrity of your queried namespaces. Implementation of rate limiting can protect against DoS/DDoS, bogus queries and PRSD attacks, and response rate limiting can mitigate reflector and amplification attacks.

DNS firewall technology is relatively new but rapidly maturing technology that can help you block malware C&C communications attempts not to mention illicit domains. You may choose to rely on a third party perhaps for a DNS firewall feed to block or modify responses to queries meeting your response policy configuration. Most firewall feed providers update their data on an hourly basis, which provides adequate protection though little in the way of zero-day attack detection or prevention.

At this stage, you should be measuring your DNS performance data, for example, queries per second, ideally by RRType, clients with the most queries (top talkers), domains with the most queries. Storing such metrics over time enables baselining of your "normal" DNS traffic characteristics. Measuring such data in near real time then affords the ability to detect any unusual traffic patterns, which could indicate an attack or other threat event. This tier level also merits more disciplined documentation of configuring and responding to events including documentation of DNSSEC signing and rollover policies, DNS firewall logging, alerting and response procedures, and traffic anomaly investigation and mitigation steps. Such documentation should be reviewed and approved with management buy-in to homogenize defensive, detection, and recovery procedures.

It's also important to consider logging your DNS transaction data, not just the counts. This may sound intimidating given your DNS servers may process thousands or millions of queries a day. At minimum, you should be monitoring query rates and DNS server vitals to assure availability of DNS services and suitable performance. This information can be helpful for detecting query swells, which could indicate a service denial, tunneling, or malware related attack. While query rate monitoring serves as a possible indicator, without query details you may be at a loss to diagnose whether this situation is an attack and if so, of what sort.

Logging queries and storing history even for a brief period of time can help provide you solid data from which you can diagnose potential attacks in progress. If nothing else, DNS query data can be useful for forensic analysis. As DNS server vendor implementations evolve to simplify this process, with ISC BIND natively supporting dnstap for example in its 9.11 release, the configuration of logging is simplified. You still need to provision a logging aggregator and collector such as Splunk (95) for generic logging collection or Stratos (96) for DNS-specific logging from which searches and queries may be made when researching threat events.

Tier 3 raises the level of DNS security sophistication with the manual or automated analysis of DNS packets. This analysis should consider individual packets as well as multipacket exchanges between a client resolver and server or across several clients. Such analysis can be useful to detect DNS tunneling, fast or double flux, DGAs, and related malware communications techniques. This process may start as reporting alone, where packet analysis could highlight suspected attack vectors, enabling a DNS security analyst to apply judgment to continue monitoring or take evasive action. Automation of such analysis can reduce the window of exposure from the onslaught of an attack to the imposition of defensive controls.

"Adaptive" is a fitting term for Tier 4 as defined by the NIST Cybersecurity Framework. This level builds on underlying configuration and analytics to correlate DNS transactions to automate threat detection and mitigation while minimizing false positives. Broader network security systems should integrate and leverage DNS security data to support greater network visibility and intelligence in characterizing and mitigating threat events. Evolution from manual correlation to automated processes reduces the threat event response time. Dynamic counter measures and techniques will continue to evolve for DNS allowing for faster response times to detect and mitigate attacks.

Integration of DNS events and alerts into aggregating SIEM systems will facilitate a more well-rounded decision process that can be largely automated. This is certainly a lofty vision of a multifaceted, near real time threat detection system, but to this we should strive. As attacks emerge with increasing sophistication, the "time to detect" will become a critical criterion for DNS and network security systems. Automated mitigation based on rapid detection would reduce the window of time during which an attacker may operate, though minimizing false positives would remain a key requirement. Analytics and artificial intelligence technologies at scale will help fuel the development of these products and services in the coming years.

With a secure DNS infrastructure, you can take advantage of the rich repository that DNS provides. You may secure other core applications like web browsing using DANE and email using DANE, SPF, DKIM, and DMARC. Securing voice over IP (VoIP) technologies using ENUM, a technology that relies on DNS for telephone number lookups, can improve voice services robustness. The importance of detection of malware-infested Internet devices and related security measures within the Internet of Things (IoT) recently glared for attention with massive DDoS attacks in October 2016. As connectivity proliferates, so do security vulnerabilities so security must become a built-in requirement.

There's no question that the discipline of cybersecurity will continue to gain visibility especially with the onset of recent international attacks. The Internet serves as the latest theater of battle among nations. Leveraging maturing technologies such as data science and analytics, DNS technology and supporting security systems must continue to evolve to counter ever more sophisticated attack strategies in the ongoing cybersecurity arms race.

A

CYBERSECURITY FRAMEWORK CORE DNS EXAMPLE

Subcategory	DNS Relevant Activities/Outcomes	Informative References
IDENTIFY: Asset Management (ID.AM): The data, personnel, devices, systems, and facilities that enable the organization to achieve business purposes are identified and managed consistent with their relative importance to business objectives and the organization's risk strategy.		
ID.AM-1: Physical devices and systems within the organization are inventoried	• Document inventory of recursive DNS servers • Document inventory of internal authoritative DNS servers • Document inventory of external authoritative DNS servers • Document use of external DNS providers for recursion or authoritative DNS • Document inventory of resolvers (clients, stub resolvers)	CCS CSC 1 COBIT 5 BAI09.01, BAI09.02 ISA 62443-2-1:2009 4.2.3.4 ISA 62443-3-3:2013 SR 7.8 ISO/IEC 27001:2013 A.8.1.1, A.8.1.2 NIST SP 800-53 Rev. 4 CM-8

DNS Security Management, First Edition. Michael Dooley and Timothy Rooney.
© 2017 by The Institute of Electrical and Electronic Engineers, Inc. Published 2017 by John Wiley & Sons, Inc.

Subcategory	DNS Relevant Activities/Outcomes	Informative References
ID.AM-2: Software platforms and applications within the organization are inventoried	• Document DNS vendor and software version for recursive DNS servers • Document DNS vendor and software version for authoritative DNS servers • Document DNS vendor and software version for external authoritative DNS servers • Document resolver vendor and version of resolvers (clients, stub resolvers)	CCS CSC 2 COBIT 5 BAI09.01, BAI09.02, BAI09.05 ISA 62443-2-1:2009 4.2.3.4 ISA 62443-3-3:2013 SR 7.8 ISO/IEC 27001:2013 A.8.1.1, A.8.1.2 NIST SP 800-53 Rev. 4 CM-8
ID.AM-3: Organizational communication and data flows are mapped	• Map Internet DNS resolution data flows – resolvers perhaps scoped within a given geography resolve to a given set of recursive servers which may forward to centralized caching servers and out to the Internet and back • Map internal DNS resolution data flows – resolvers perhaps scoped within a given geography resolve to a given set of recursive servers which may forward to centralized caching servers and to internal authoritative DNS servers and back • External DNS servers are queried only from Internet sources and should have recursion disabled • Map communication flows for threat monitoring, detection, reporting, escalation, recovery, postmortem, and external communications	CCS CSC 1 COBIT 5 DSS05.02 ISA 62443-2-1:2009 4.2.3.4 ISO/IEC 27001:2013 A.13.2.1 NIST SP 800-53 Rev. 4 AC-4, CA-3, CA-9, PL-8
ID.AM-4: External information systems are catalogued	• Documentation of DNS policies consistent with enterprise security policy related to external devices and services • External DNS system services for recursive and authoritative services comply with organizational and relevant regulatory security requirements • External DNS services administrator access policies including access recovery procedures are documented	COBIT 5 APO02.02 ISO/IEC 27001:2013 A.11.2.6 NIST SP 800-53 Rev. 4 AC-20, SA-9

Subcategory	DNS Relevant Activities/Outcomes	Informative References
ID.AM-5: Resources (e.g., hardware, devices, data, and software) are prioritized based on their classification, criticality, and business value	• Categorize and prioritize DNS servers accordingly, plan for contingencies where compromise could affect ability to resolve DNS (recursive, resolver) and redirect external end users (authoritative compromise) • Categorize DNS software vendor supply chain risk • Define contingencies for DNS server failure or failure of external DNS provider(s)	COBIT 5 APO03.03, APO03.04, BAI09.02 ISA 62443-2-1:2009 4.2.3.6 ISO/IEC 27001:2013 A.8.2.1 NIST SP 800-53 Rev. 4 CP-2, RA-2, SA-14
ID.AM-6: Cybersecurity roles and responsibilities for the entire workforce and third-party stakeholders (e.g., suppliers, customers, partners) are established	• Document roles and responsibilities for those responsible for each DNS server including third parties and service provider(s) • Consider DNS within business process definitions for information security and resulting risk, along with information protection needs (e.g., DNS tunneling, APTs)	COBIT 5 APO01.02, DSS06.03 ISA 62443-2-1:2009 4.3.2.3.3 ISO/IEC 27001:2013 A.6.1.1 NIST SP 800-53 Rev. 4 CP-2, PS-7, PM-11
IDENTIFY: Business Environment (ID.BE): The organization's mission, objectives, stakeholders, and activities are understood and prioritized; this information is used to inform cybersecurity roles, responsibilities, and risk management decisions.		
ID.BE-1: The organization's role in the supply chain is identified and communicated	• Procure and deploy DNS servers and/or DNS services which include required security features, noncustom configurations, from diverse suppliers on an approved vendor list from approved countries	COBIT 5 APO08.04, APO08.05, APO10.03, APO10.04, APO10.05 ISO/IEC 27001:2013 A.15.1.3, A.15.2.1, A.15.2.2 NIST SP 800-53 Rev. 4 CP-2, SA-12
ID.BE-2: The organization's place in critical infrastructure and its industry sector is identified and communicated	• Categorize and prioritize DNS components according to criticality, plan for contingencies when compromise could affect ability to resolve DNS (forwarder, recursive, referrer, resolver) and redirect external end users (authoritative compromise) • Define contingencies for DNS server failure or failure of external DNS provider(s)	COBIT 5 APO02.06, APO03.01 NIST SP 800-53 Rev. 4 PM-8

Subcategory	DNS Relevant Activities/Outcomes	Informative References
ID.BE-3: Priorities for organizational mission, objectives, and activities are established and communicated	• Consider DNS within business process definitions for information security and resulting risk, criticality assessment, along with information protection needs	COBIT 5 APO02.01, APO02.06, APO03.01 ISA 62443-2-1:2009 4.2.2.1, 4.2.3.6 NIST SP 800-53 Rev. 4 PM-11, SA-14
ID.BE-4: Dependencies and critical functions for delivery of critical services are established	• Include DNS infrastructure in critical infrastructure plan • Plan for protected and/or uninterruptible power for DNS servers as appropriate • Plan for diverse telecom facilities to the Internet and DNS service providers • Plan for diversity for external trust sector with in-house and/or one or more external DNS service providers • Consider DNS within business process definitions for information security and resulting risk, criticality assessment, along with information protection needs	ISO/IEC 27001:2013 A.11.2.2, A.11.2.3, A.12.1.3 NIST SP 800-53 Rev. 4 CP-8, PE-9, PE-11, PM-8, SA-14
ID.BE-5: Resilience requirements to support delivery of critical services are established	• Procure and deploy DNS servers and/or DNS services which include security features, noncustom configurations, from diverse suppliers on an approved vendor list from approved countries • Deploy DNS servers in accordance with trust sector architecture • Deploy DNS servers with both IPv4 and IPv6 communications capabilities • External DNS services are contracted to include high availability SLAs with one or more providers • Consider DNS within business process definitions for information security and resulting risk, criticality assessment, along with information protection needs	COBIT 5 DSS04.02 ISO/IEC 27001:2013 A.11.1.4, A.17.1.1, A.17.1.2, A.17.2.1 NIST SP 800-53 Rev. 4 CP-2, CP-11, SA-14

Subcategory	DNS Relevant Activities/Outcomes	Informative References
IDENTIFY: Governance (ID.GV): The policies, procedures, and processes to manage and monitor the organization's regulatory, legal, risk, environmental, and operational requirements are understood and inform the management of cybersecurity risk.		
ID.GV-1: Organizational information security policy is established	• DNS inclusion in the organization's information security policy document, which is approved by management and published to employees and relevant external associates.	COBIT 5 APO01.03, EDM01.01, EDM01.02 ISA 62443-2-1:2009 4.3.2.6 ISO/IEC 27001:2013 A.5.1.1 NIST SP 800-53 Rev. 4 -1 controls from all families
ID.GV-2: Information security roles and responsibilities are coordinated and aligned with internal roles and external partners	• Roles and responsibilities for DNS security functions and processes and documented and communicated to associated employees and/or external (third-party) associates	COBIT 5 APO13.12 ISA 62443-2-1:2009 4.3.2.3.3 ISO/IEC 27001:2013 A.6.1.1, A.7.2.1 NIST SP 800-53 Rev. 4 PM-1, PS-7
ID.GV-3: Legal and regulatory requirements regarding cybersecurity, including privacy, and civil liberties, obligations, are understood and managed	• DNS inclusion in the organization's information security policy document, which is approved by management and published to employees and relevant external associates. Training is provided and acknowledgement of understanding of material is obtained	COBIT 5 MEA03.01, MEA03.04 ISA 62443-2-1:2009 4.4.3.7 ISO/IEC 27001:2013 A.18.1 NIST SP 800-53 Rev. 4 -1 controls from all families (except PM-1)
ID.GV-4: Governance and risk management processes address cybersecurity risks	• DNS inclusion in the organization's risk management processes which is approved by management and published to employees and relevant external associates. Training is provided and acknowledgement of understanding of material is obtained	COBIT 5 DSS04.02 ISA 62443-2-1:2009 4.2.3.1, 4.2.3.3, 4.2.3.8, 4.2.3.9, 4.2.3.11, 4.3.2.4.3, 4.3.2.6.3 NIST SP 800-53 Rev. 4 PM-9, PM-11

Subcategory	DNS Relevant Activities/Outcomes	Informative References
IDENTIFY: Risk Assessment (ID.RA): The organization understands the cybersecurity risk to organizational operations (including mission, functions, image, or reputation), organizational assets, and individuals.		
ID.RA-1: Asset vulnerabilities are identified and documented	• Document and track vulnerabilities for your DNS server hardware components • Document and track vulnerabilities for your DNS server kernels • Document and track vulnerabilities for your DNS server operating systems • Document and track vulnerabilities for your DNS server software applications • Document and track vulnerabilities for DNS resolver software applications on end user and other devices • Document and track vulnerabilities for the DNS protocol	CCS CSC 4 COBIT 5 APO12.01, APO12.02, APO12.03, APO12.04 ISA 62443-2-1:2009 4.2.3, 4.2.3.7, 4.2.3.9, 4.2.3.12 ISO/IEC 27001:2013 A.12.6.1, A.18.2.3 NIST SP 800-53 Rev. 4 CA-2, CA-7, CA-8, RA-3, RA-5, SA-5, SA-11, SI-2, SI-4, SI-5
ID.RA-2: Threat and vulnerability information is received from information sharing forums and sources	• Actively monitor threat and vulnerability information sources like CERT, your DNS software vendor alerts and related reputable web resources • Collaborate with other security personnel especially within your industry to promote open sharing of threats, vulnerabilities, and mitigations • Subscribe to security feeds offered by vendors who produce the operating systems and applications (DNS) software in use within your network	ISA 62443-2-1:2009 4.2.3, 4.2.3.9, 4.2.3.12 ISO/IEC 27001:2013 A.6.1.4 NIST SP 800-53 Rev. 4 PM-15, PM-16, SI-5
ID.RA-3: Threats, both internal and external, are identified and documented	• Monitor, document, and track detected and reported vulnerabilities for DNS resolver and server software applications • Document threats related to personnel issues, human error, as well as natural and unnatural disasters • Monitor, document, and track detected and reported vulnerabilities for external DNS services and applications	COBIT 5 APO12.01, APO12.02, APO12.03, APO12.04 ISA 62443-2-1:2009 4.2.3, 4.2.3.9, 4.2.3.12 NIST SP 800-53 Rev. 4 RA-3, SI-5, PM-12, PM-16

Subcategory	DNS Relevant Activities/Outcomes	Informative References
ID.RA-4: Potential business impacts and likelihoods are identified	• For each identified threat and vulnerability, document the related business impacts should the threat materialize • For each identified threat and vulnerability, document the likelihood of the occurrence	COBIT 5 DSS04.02 ISA 62443-2-1:2009 4.2.3, 4.2.3.9, 4.2.3.12 NIST SP 800-53 Rev. 4 RA-2, RA-3, PM-9, PM-11, SA-14
ID.RA-5: Threats, vulnerabilities, likelihoods, and impacts are used to determine risk	• Assess the risk of each identified threat and vulnerability based on the likelihood and related business impacts of the occurrence • Iterate your analysis to consider the relative risk for higher priority assets, e.g., DNS servers, as identified in subcategory ID.AM-5. A given threat may have a higher impact when imposed on a master DNS server versus a slave, for example	COBIT 5 APO12.02 ISO/IEC 27001:2013 A.12.6.1 NIST SP 800-53 Rev. 4 RA-2, RA-3, PM-16
ID.RA-6: Risk responses are identified and prioritized	• Define and document the response procedures for each risk identified	COBIT 5 APO12.05, APO13.02 NIST SP 800-53 Rev. 4 PM-4, PM-9
IDENTIFY: Risk Management Strategy (ID.RM): The organization's priorities, constraints, risk tolerances, and assumptions are established and used to support operational risk decisions.		
ID.RM-1: Risk management processes are established, managed, and agreed to by organizational stakeholders	• Assess the risk of each identified threat and vulnerability based on the likelihood and related business impacts of the occurrence	COBIT 5 APO12.04, APO12.05, APO13.02, BAI02.03, BAI04.02 ISA 62443-2-1:2009 4.3.4.2 NIST SP 800-53 Rev. 4 PM-9
ID.RM-2: Organizational risk tolerance is determined and clearly expressed	• Define and gain consensus regarding the process for determining and documenting organizational risk including DNS	COBIT 5 APO12.06 ISA 62443-2-1:2009 4.3.2.6.5 NIST SP 800-53 Rev. 4 PM-9

Subcategory	DNS Relevant Activities/Outcomes	Informative References
ID.RM-3: The organization's determination of risk tolerance is informed by its role in critical infrastructure and sector specific risk analysis	• Define and gain consensus regarding risk tolerance for each identified DNS risk	NIST SP 800-53 Rev. 4 PM-8, PM-9, PM-11, SA-14
PROTECT: Access Control (PR.AC): Access to assets and associated facilities is limited to authorized users, processes, or devices, and to authorized activities and transactions.		
PR.AC-1: Identities and credentials are managed for authorized devices and users	• For each DNS component, identify and document authorized users • For each DNS component, enforce credentials requirements • For each DNS component, manage credentials refresh policies • For each DNS component, document an approval process for authorizing new users or expansion or contraction of existing users' permissions • For each DNS component, audit who has access and respective level of access and confirm or modify based on current job role	CCS CSC 16 COBIT 5 DSS05.04, DSS06.03 ISA 62443-2-1:2009 4.3.3.5.1 ISA 62443-3-3:2013 SR 1.1, SR 1.2, SR 1.3, SR 1.4, SR 1.5, SR 1.7, SR 1.8, SR 1.9 ISO/IEC 27001:2013 A.9.2.1, A.9.2.2, A.9.2.4, A.9.3.1, A.9.4.2, A.9.4.3 NIST SP 800-53 Rev. 4 AC-2, IA Family
PR.AC-2: Physical access to assets is managed and protected	• For each DNS component, define appropriate physical access protection requirements • For each DNS component, deploy in accordance with physical access requirements, e.g., within badge accessible datacenter • Document physical access permissions to restricted areas where DNS servers are deployed to confirm user permission with user appropriateness for access • Audit physical access logs and surveillance videos if appropriate with respect to access to restricted areas to confirm only appropriate users access	COBIT 5 DSS01.04, DSS05.05 ISA 62443-2-1:2009 4.3.3.3.2, 4.3.3.3.8 ISO/IEC 27001:2013 A.11.1.1, A.11.1.2, A.11.1.4, A.11.1.6, A.11.2.3 NIST SP 800-53 Rev. 4 PE-2, PE-3, PE-4, PE-5, PE-6, PE-9

Subcategory	DNS Relevant Activities/Outcomes	Informative References
PR.AC-3: Remote access is managed	• Deploy an extranet trust sector to partition remote access from a DNS perspective • Only staff (internal or contractors) with a job function and scope of responsibility necessitating remote access are permitted remote access • Login/password authentication is required and communications should require an encrypted connection • Command sets are restricted if possible to only those required by each user	COBIT 5 APO13.01, DSS01.04, DSS05.03 ISA 62443-2-1:2009 4.3.3.6.6 ISA 62443-3-3:2013 SR 1.13, SR 2.6 ISO/IEC 27001:2013 A.6.2.2, A.13.1.1, A.13.2.1 NIST SP 800-53 Rev. 4 AC-17, AC-19, AC-20
PR.AC-4: Access permissions are managed, incorporating the principles of least privilege and separation of duties	• For each user identity, the level of access should provide only those commands or functions required of that user to perform his or her job function	CCS CSC 12, 15 ISA 62443-2-1:2009 4.3.3.7.3 ISA 62443-3-3:2013 SR 2.1 ISO/IEC 27001:2013 A.6.1.2, A.9.1.2, A.9.2.3, A.9.4.1, A.9.4.4 NIST SP 800-53 Rev. 4 AC-2, AC-3, AC-5, AC-6, AC-16
PR.AC-5: Network integrity is protected, incorporating network segregation where appropriate	• Deployment of DNS components in accordance with defined trust sectors provides containment to the corresponding trust sector	ISA 62443-2-1:2009 4.3.3.4 ISA 62443-3-3:2013 SR 3.1, SR 3.8 ISO/IEC 27001:2013 A.13.1.1, A.13.1.3, A.13.2.1 NIST SP 800-53 Rev. 4 AC-4, SC-7
PROTECT: Awareness and Training (PR.AT): The organization's personnel and partners are provided cybersecurity awareness education and are adequately trained to perform their information security related duties and responsibilities consistent with related policies, procedures, and agreements.		
PR.AT-1: All users are informed and trained	• Devise, develop, and deliver a training program for IT security including DNS elements and garner acknowledgement of key aspects • Keep training material up to date and require periodic update training for all users	CCS CSC 9 COBIT 5 APO07.03, BAI05.07 ISA 62443-2-1:2009 4.3.2.4.2 ISO/IEC 27001:2013 A.7.2.2 NIST SP 800-53 Rev. 4 AT-2, PM-13

Subcategory	DNS Relevant Activities/Outcomes	Informative References
PR.AT-2: Privileged users understand roles and responsibilities	• Document well-defined job descriptions outlining roles and responsibilities for all users • For users whose job functions require privileged access, provide training and garner acknowledgement of respective responsibilities	CCS CSC 9 COBIT 5 APO07.02, DSS06.03 ISA 62443-2-1:2009 4.3.2.4.2, 4.3.2.4.3 ISO/IEC 27001:2013 A.6.1.1, A.7.2.2 NIST SP 800-53 Rev. 4 AT-3, PM-13
PR.AT-3: Third-party stakeholders (e.g., suppliers, customers, partners) understand roles and responsibilities	• For third-party stakeholders who require access, provide training and garner acknowledgement of respective responsibilities • For external DNS services, document respective roles and responsibilities	CCS CSC 9 COBIT 5 APO07.03, APO10.04, APO10.05 ISA 62443-2-1:2009 4.3.2.4.2 ISO/IEC 27001:2013 A.6.1.1, A.7.2.2 NIST SP 800-53 Rev. 4 PS-7, SA-9
PR.AT-4: Senior executives understand roles and responsibilities	• For executives, provide training and garner acknowledgement of respective roles and responsibilities	CCS CSC 9 COBIT 5 APO07.03 ISA 62443-2-1:2009 4.3.2.4.2 ISO/IEC 27001:2013 A.6.1.1, A.7.2.2, NIST SP 800-53 Rev. 4 AT-3, PM-13
PR.AT-5: Physical and information security personnel understand roles and responsibilities	• For physical and information security personnel, provide training and garner acknowledgement of respective roles and responsibilities	CCS CSC 9 COBIT 5 APO07.03 ISA 62443-2-1:2009 4.3.2.4.2 ISO/IEC 27001:2013 A.6.1.1, A.7.2.2, NIST SP 800-53 Rev. 4 AT-3, PM-13
PROTECT: Data Security (PR.DS): Information and records (data) are managed consistent with the organization's risk strategy to protect the confidentiality, integrity, and availability of information.		
PR.DS-1: Data at rest is protected	• Secure DNS component hardware, harden your operating system, kernel, and software • ACLs and transaction keys are implemented to protect authoritative DNS data from updates and zone transfers	CCS CSC 17 COBIT 5 APO01.06, BAI02.01, BAI06.01, DSS06.06 ISA 62443-3-3:2013 SR 3.4, SR 4.1

Subcategory	DNS Relevant Activities/Outcomes	Informative References
	• ACLs are defined to control management access and management transactions are encrypted • Sign DNS zone data using DNSSEC; secure DNSSEC private keys and document rollover processes including emergency rollover • Periodic backups and storage of DNS configuration data offsite provides a fallback for restoration in the event of server failure • DNS configuration and logging data stored offsite must be transported and stored securely via encryption • Monitor vulnerability sources and deploy vendor patches affecting data at rest • Periodically audit DNS component access logs and functions performed to align with roles and responsibilities	ISO/IEC 27001:2013 A.8.2.3 NIST SP 800-53 Rev. 4 SC-28
PR.DS-2: Data in transit is protected	• Configuration and zone data updated via console, remote access, or IPAM system are authenticated and encrypted • DNS resolution data is signed using DNSSEC. Authoritative zone data is signed and recursive servers perform DNSSEC validation • Source port and TXID numbers are randomized for queries • Query case is randomized • ACLs are configured to control entitlement for queries, cache access, and zone transfers • Zone transfers and notify's are signed • Secure the resolver-to-recursive DNS server link with cookies or DNSCrypt • Monitor for DNS tunneling • Implement a DNS firewall and monitor for malware C&C queries • Monitor vulnerability sources and deploy vendor patches affecting DNS data in motion • Implement DoS/DDoS controls such as inbound and outbound rate limiting, query throttling, and anycast • Periodically audit DNS component access logs and functions performed to align with roles and responsibilities	CCS CSC 17 COBIT 5 APO01.06, DSS06.06 ISA 62443-3-3:2013 SR 3.1, SR 3.8, SR 4.1, SR 4.2 ISO/IEC 27001:2013 A.8.2.3, A.13.1.1, A.13.2.1, A.13.2.3, A.14.1.2, A.14.1.3 NIST SP 800-53 Rev. 4 SC-8

Subcategory	DNS Relevant Activities/Outcomes	Informative References
PR.DS-3: Assets are formally managed throughout removal, transfers, and disposition	• For all DNS servers and resolvers, any proposed addition, movement, or removal of DNS components must be documented, reviewed, and approved by involved parties prior to commencement • Additions, changes, or deletions of DNS service provider parameters are carefully managed, tracked, and verified	COBIT 5 BAI09.03 ISA 62443-2-1:2009 4. 4.3.3.3.9, 4.3.4.4.1 ISA 62443-3-3:2013 SR 4.2 ISO/IEC 27001:2013 A.8.2.3, A.8.3.1, A.8.3.2, A.8.3.3, A.11.2.7 NIST SP 800-53 Rev. 4 CM-8, MP-6, PE-16
PR.DS-4: Adequate capacity to ensure availability is maintained	• Deploy sufficient DNS capacity within each trust sector to provide acceptable resolution performance even in the event of an outage within the network and/or within your DNS server infrastructure • Monitor capacity utilization over time and deploy additional infrastructure as necessitated by growing demand as appropriate	COBIT 5 APO13.01 ISA 62443-3-3:2013 SR 7.1, SR 7.2 ISO/IEC 27001:2013 A.12.3.1 NIST SP 800-53 Rev. 4 AU-4, CP-2, SC-5
PR.DS-5: Protections against data leaks are implemented	• ACLs are implemented to protect access to internal authoritative DNS data resolution • External DNS authoritative data consists only of data relevant to externally accessible services as published on in-house or service provider DNS • DNS tunneling detection and mitigation strategies are in place • DNS firewall configuration helps prevent access to external sites for possible exfiltration	CCS CSC 17 COBIT 5 APO01.06 ISA 62443-3-3:2013 SR 5.2 ISO/IEC 27001:2013 A.6.1.2, A.7.1.1, A.7.1.2, A.7.3.1, A.8.2.2, A.8.2.3, A.9.1.1, A.9.1.2, A.9.2.3, A.9.4.1, A.9.4.4, A.9.4.5, A.13.1.3, A.13.2.1, A.13.2.3, A.13.2.4, A.14.1.2, A.14.1.3 NIST SP 800-53 Rev. 4 AC-4, AC-5, AC-6, PE-19, PS-3, PS-6, SC-7, SC-8, SC-13, SC-31, SI-4
PR.DS-6: Integrity checking mechanisms are used to verify software, firmware, and information integrity	• DNS configuration and zone file integrity checks are implemented to verify successful transfer and to detect changes • DNSSEC validation provides DNS resolution data integrity checking and origin authentication as well as authenticated denial of existence	ISA 62443-3-3:2013 SR 3.1, SR 3.3, SR 3.4, SR 3.8 ISO/IEC 27001:2013 A.12.2.1, A.12.5.1, A.14.1.2, A.14.1.3 NIST SP 800-53 Rev. 4 SI-7

Subcategory	DNS Relevant Activities/Outcomes	Informative References
PR.DS-7: The development and testing environment(s) are separate from the production environment	• DNS resolvers and servers are deployed within a lab network separate from the production network for testing of new releases, patches, and new features	COBIT 5 BAI07.04 ISO/IEC 27001:2013 A.12.1.4 NIST SP 800-53 Rev. 4 CM-2
PROTECT: Information Protection Processes and Procedures (PR.IP): Security policies (that address purpose, scope, roles, responsibilities, management commitment, and coordination among organizational entities), processes, and procedures are maintained and used to manage protection of information systems and assets.		
PR.IP-1: A baseline configuration of information technology/industrial control systems is created and maintained	• For end user systems, the resolver software is defined as included with the build, which almost always consists of that supplied with the corresponding device operating system • For DNS servers, the list of hardware and software components, including operating systems, DNS software, security, monitoring and auditing utilities and so on, is documented	CCS CSC 3, 10 COBIT 5 BAI10.01, BAI10.02, BAI10.03, BAI10.05 ISA 62443-2-1:2009 4.3.4.3.2, 4.3.4.3.3 ISA 62443-3-3:2013 SR 7.6 ISO/IEC 27001:2013 A.12.1.2, A.12.5.1, A.12.6.2, A.14.2.2, A.14.2.3, A.14.2.4 NIST SP 800-53 Rev. 4 CM-2, CM-3, CM-4, CM-5, CM-6, CM-7, CM-9, SA-10
PR.IP-2: A System Development Life Cycle to manage systems is implemented	• Apply security engineering principles to your IT network and systems through integration with your systems development lifecycle, including in-house development projects from requirements analysis through coding and testing; product procurement activities likewise should include security considerations	COBIT 5 APO13.01 ISA 62443-2-1:2009 4.3.4.3.3 ISO/IEC 27001:2013 A.6.1.5, A.14.1.1, A.14.2.1, A.14.2.5 NIST SP 800-53 Rev. 4 SA-3, SA-4, SA-8, SA-10, SA-11, SA-12, SA-15, SA-17, PL-8

Subcategory	DNS Relevant Activities/Outcomes	Informative References
PR.IP-3: Configuration change control processes are in place	• The baseline configuration is a controlled document, meaning that any additions, changes, or deletions are proposed, reviewed, approved, and communicated among relevant parties • Changes to the configuration of the authorized software, e.g., the DNS configuration, should also be planned, reviewed, approved, and staged	COBIT 5 BAI06.01, BAI01.06 ISA 62443-2-1:2009 4.3.4.3.2, 4.3.4.3.3 ISA 62443-3-3:2013 SR 7.6 ISO/IEC 27001:2013 A.12.1.2, A.12.5.1, A.12.6.2, A.14.2.2, A.14.2.3, A.14.2.4 NIST SP 800-53 Rev. 4 CM-3, CM-4, SA-10
PR.IP-4: Backups of information are conducted, maintained, and tested periodically	• Backup DNS configuration and zone repositories regularly for each set of servers as well as audit logs	COBIT 5 APO13.01 ISA 62443-2-1:2009 4.3.4.3.9 ISA 62443-3-3:2013 SR 7.3, SR 7.4 ISO/IEC 27001:2013 A.12.3.1, A.17.1.2A.17.1.3, A.18.1.3 NIST SP 800-53 Rev. 4 CP-4, CP-6, CP-9
PR.IP-5: Policy and regulations regarding the physical operating environment for organizational assets are met	• Document and maintain policies and regulations regarding the physical operating environment for organizational assets such as DNS servers. This includes providing emergency power shut-off, fire protection, temperature and humidity controls, water damage protection, and server room or datacenter access controls and auditing.	COBIT 5 DSS01.04, DSS05.05 ISA 62443-2-1:2009 4.3.3.3.1 4.3.3.3.2, 4.3.3.3.3, 4.3.3.3.5, 4.3.3.3.6 ISO/IEC 27001:2013 A.11.1.4, A.11.2.1, A.11.2.2, A.11.2.3 NIST SP 800-53 Rev. 4 PE-10, PE-12, PE-13, PE-14, PE-15, PE-18
PR.IP-6: Data is destroyed according to policy	• DNS configuration, resolution, and log data deemed for disposal is destroyed according to policy to prevent information theft	COBIT 5 BAI09.03 ISA 62443-2-1:2009 4.3.4.4.4 ISA 62443-3-3:2013 SR 4.2 ISO/IEC 27001:2013 A.8.2.3, A.8.3.1, A.8.3.2, A.11.2.7 NIST SP 800-53 Rev. 4 MP-6

Subcategory	DNS Relevant Activities/Outcomes	Informative References
PR.IP-7: Protection processes are continuously improved	• DNS data protection processes are continuously improved through periodic review, new technologies, and lessons learned	COBIT 5 APO11.06, DSS04.05 ISA 62443-2-1:2009 4.4.3.1, 4.4.3.2, 4.4.3.3, 4.4.3.4, 4.4.3.5, 4.4.3.6, 4.4.3.7, 4.4.3.8 NIST SP 800-53 Rev. 4 CA-2, CA-7, CP-2, IR-8, PL-2, PM-6
PR.IP-8: Effectiveness of protection technologies is shared with appropriate parties	• DNS data protection technology effectiveness is shared with appropriate parties	ISO/IEC 27001:2013 A.16.1.6 NIST SP 800-53 Rev. 4 AC-21, CA-7, SI-4
PR.IP-9: Response plans (Incident Response and Business Continuity) and recovery plans (Incident Recovery and Disaster Recovery) are in place and managed	• DNS incident response and recovery plans are documented, communicated, and managed • DNS aspects of business continuity are incorporated into business continuity and disaster recovery plans	COBIT 5 DSS04.03 ISA 62443-2-1:2009 4.3.2.5.3, 4.3.4.5.1 ISO/IEC 27001:2013 A.16.1.1, A.17.1.1, A.17.1.2 NIST SP 800-53 Rev. 4 CP-2, IR-8
PR.IP-10: Response and recovery plans are tested	• DNS incident response and recovery plans are tested and results fed back to improving such plans	ISA 62443-2-1:2009 4.3.2.5.7, 4.3.4.5.11 ISA 62443-3-3:2013 SR 3.3 ISO/IEC 27001:2013 A.17.1.3 NIST SP 800-53 Rev.4 CP-4, IR-3, PM-14
PR.IP-11: Cybersecurity is included in human resource practices (e.g., deprovisioning, personnel screening)	• Security is incorporated into human resource processes relating to recruiting, hiring, evaluations as appropriate, training, and deprovisioning	COBIT 5 APO07.01, APO07.02, APO07.03, APO07.04, APO07.05 ISA 62443-2-1:2009 4.3.3.2.1, 4.3.3.2.2, 4.3.3.2.3 ISO/IEC 27001:2013 A.7.1.1, A.7.3.1, A.8.1.4 NIST SP 800-53 Rev. 4 PS Family

Subcategory	DNS Relevant Activities/Outcomes	Informative References
PR.IP-12: A vulnerability management plan is developed and implemented	• A DNS vulnerability management plan is developed and implemented incorporating those discussed within this book	ISO/IEC 27001:2013 A.12.6.1, A.18.2.2 NIST SP 800-53 Rev. 4 RA-3, RA-5, SI-2
PROTECT: Maintenance (PR.MA): Maintenance and repairs of industrial control and information system components is performed consistent with policies and procedures.		
PR.MA-1: Maintenance and repair of organizational assets is performed and logged in a timely manner, with approved and controlled tools	• Maintenance and repair of DNS servers is performed and logged in a timely manner, with approved and controlled tools • Maintenance and repairs uses pre-approved documented tools and processes • Any repairs requiring support from outside personnel, e.g., vendor staff, is pre-authorized and outside personnel should sign in and out and be escorted at all times by an authorized organization team member • Repairs requiring removal of a DNS component are approved along with associated contingencies and the component sanitized of any sensitive information such as user accounts	COBIT 5 BAI09.03 ISA 62443-2-1:2009 4.3.3.3.7 ISO/IEC 27001:2013 A.11.1.2, A.11.2.4, A.11.2.5 NIST SP 800-53 Rev. 4 MA-2, MA-3, MA-5
PR.MA-2: Remote maintenance of organizational assets is approved, logged, and performed in a manner that prevents unauthorized access	• Any DNS server maintenance or repair performed remotely must be pre-approved and a remote connection opened for the duration of the activity • Strong credentials for remote access are needed for identity verification • Remote access is to be logged as is DNS server commands and diagnostic actions	COBIT 5 DSS05.04 ISA 62443-2-1:2009 4.3.3.6.5, 4.3.3.6.6, 4.3.3.6.7, 4.4.4.6.8 ISO/IEC 27001:2013 A.11.2.4, A.15.1.1, A.15.2.1 NIST SP 800-53 Rev. 4 MA-4

Subcategory	DNS Relevant Activities/Outcomes	Informative References
PROTECT: Protective Technology (PR.PT): Technical security solutions are managed to ensure the security and resilience of systems and assets, consistent with related policies, procedures, and agreements.		
PR.PT-1: Audit/log records are determined, documented, implemented, and reviewed in accordance with policy	• DNS server and management system logs are identified and configured for tracking and storage for documentation of log records	CCS CSC 14 COBIT 5 APO11.04 ISA 62443-2-1:2009 4.3.3.3.9, 4.3.3.5.8, 4.3.4.4.7, 4.4.2.1, 4.4.2.2, 4.4.2.4 ISA 62443-3-3:2013 SR 2.8, SR 2.9, SR 2.10, SR 2.11, SR 2.12 ISO/IEC 27001:2013 A.12.4.1, A.12.4.2, A.12.4.3, A.12.4.4, A.12.7.1 NIST SP 800-53 Rev. 4 AU Family
PR.PT-2: Removable media is protected and its use restricted according to policy	• Any removable media must be protected and its use restricted according to policy. Vendors may supply software updates via USB, DVD, or other removable media format. Configuration and backup information may also be copied to removable media for backups • Swappable hard drives may contain configuration or sensitive information. Such media must be securely stored to prevent unauthorized access to the media and such media must be securely transported if necessary and must be sanitized prior to disposal by the removal of any sensitive information stored on the media	COBIT 5 DSS05.02, APO13.01 ISA 62443-3-3:2013 SR 2.3 ISO/IEC 27001:2013 A.8.2.2, A.8.2.3, A.8.3.1, A.8.3.3, A.11.2.9 NIST SP 800-53 Rev. 4 MP-2, MP-4, MP-5, MP-7

Subcategory	DNS Relevant Activities/Outcomes	Informative References
PR.PT-3: Access to systems and assets is controlled, incorporating the principle of least functionality	• Only those users whose job function and role necessitate access are provided access to a given DNS component • The breadth of functionality permitted for each user is constrained to the extent possible by permission controls on the device • DNS component access logs are periodically audited to verify appropriate user access and functions executed match job function • The functionality of the device itself is constrained to the minimum functionality required to perform that device's role. Hence, for a DNS server, any non-DNS related services excepting those necessary for diagnostics and auditing are removed. In addition, restrictions are defined for unnecessary TCP/UDP ports, system or application files, processes, users, and the file system. DNS configuration maps to the servers' respective trust sector	COBIT 5 DSS05.02 ISA 62443-2-1:2009 4.3.3.5.1, 4.3.3.5.2, 4.3.3.5.3, 4.3.3.5.4, 4.3.3.5.5, 4.3.3.5.6, 4.3.3.5.7, 4.3.3.5.8, 4.3.3.6.1, 4.3.3.6.2, 4.3.3.6.3, 4.3.3.6.4, 4.3.3.6.5, 4.3.3.6.6, 4.3.3.6.7, 4.3.3.6.8, 4.3.3.6.9, 4.3.3.7.1, 4.3.3.7.2, 4.3.3.7.3, 4.3.3.7.4 ISA 62443-3-3:2013 SR 1.1, SR 1.2, SR 1.3, SR 1.4, SR 1.5, SR 1.6, SR 1.7, SR 1.8, SR 1.9, SR 1.10, SR 1.11, SR 1.12, SR 1.13, SR 2.1, SR 2.2, SR 2.3, SR 2.4, SR 2.5, SR 2.6, SR 2.7 ISO/IEC 27001:2013 A.9.1.2 NIST SP 800-53 Rev. 4 AC-3, CM-7
PR.PT-4: Communications and control networks are protected	• The deployment of trust sectors with accompanying ACLs and associated trust sector controls provide defense in depth protection • Management network access to DNS components is authenticated and encrypted	CCS CSC 7 COBIT 5 DSS05.02, APO13.01 ISA 62443-3-3:2013 SR 3.1, SR 3.5, SR 3.8, SR 4.1, SR 4.3, SR 5.1, SR 5.2, SR 5.3, SR 7.1, SR 7.6 ISO/IEC 27001:2013 A.13.1.1, A.13.2.1 NIST SP 800-53 Rev. 4 AC-4, AC-17, AC-18, CP-8, SC-7

Subcategory	DNS Relevant Activities/Outcomes	Informative References
DETECT: Anomalies and Events (DE.AE): Anomalous activity is detected in a timely manner and the potential impact of events is understood.		
DE.AE-1: A baseline of network operations and expected data flows for users and systems is established and managed	• DNS traffic is monitored and tracked historically to define a baseline • DNS server vitals (CPU, HD, memory, I/O) are monitored and tracked historically to define a baseline	COBIT 5 DSS03.01 ISA 62443-2-1:2009 4.4.3.3 NIST SP 800-53 Rev. 4 AC-4, CA-3, CM-2, SI-4
DE.AE-2: Detected events are analyzed to understand attack targets and methods	• DNS traffic anomalies are analyzed to characterize each as a possible attack or otherwise; if an attack, attack details are discovered and documented for tracking and comparison with similarities to prior attacks to possibly apply prior solutions based on past lessons learned	ISA 62443-2-1:2009 4.3.4.5.6, 4.3.4.5.7, 4.3.4.5.8 ISA 62443-3-3:2013 SR 2.8, SR 2.9, SR 2.10, SR 2.11, SR 2.12, SR 3.9, SR 6.1, SR 6.2 ISO/IEC 27001:2013 A.16.1.1, A.16.1.4 NIST SP 800-53 Rev. 4 AU-6, CA-7, IR-4, SI-4
DE.AE-3: Event data are aggregated and correlated from multiple sources and sensors	• DNS event data is aggregated and correlated with relevant network data from network and server event monitoring systems to supplement attack characterization and breadth • DNS event data is securely transmitted to broader network security event information systems to support correlation and troubleshooting	ISA 62443-3-3:2013 SR 6.1 NIST SP 800-53 Rev. 4 AU-6, CA-7, IR-4, IR-5, IR-8, SI-4
DE.AE-4: Impact of events is determined	• The impact of DNS events is determined to facilitate prioritization of mitigation efforts	COBIT 5 APO12.06 NIST SP 800-53 Rev. 4 CP-2, IR-4, RA-3, SI-4

Subcategory	DNS Relevant Activities/Outcomes	Informative References
DE.AE-5: Incident alert thresholds are established	• An incident response plan is established to define potential incidents and the corresponding response plan • The incident response plan is reviewed, approved, and updated periodically • Thresholds and alerts in monitoring systems are established to detect and report potential security incidents, e.g., for process and hardware states and well as I/O volumes. The response plan defines incident analysis, containment, eradication, and recovery, along with the communication of status reporting	COBIT 5 APO12.06 ISA 62443-2-1:2009 4.2.3.10 NIST SP 800-53 Rev. 4 IR-4, IR-5, IR-8
DETECT: Security Continuous Monitoring (DE.CM): The information system and assets are monitored at discrete intervals to identify cybersecurity events and verify the effectiveness of protective measures.		
DE.CM-1: The network is monitored to detect potential cybersecurity events	• DNS traffic is monitored to detect cybersecurity events including general and DNS traffic volume (DoS/DDoS, PRSD, reflector), NXDOMAINS (bogus queries, DGAs, malware/APTs), and unusual traffic patterns (tunneling, malware/APTs) • DNS server vitals (CPU, HD, memory, I/O) are monitored to detect unusually high utilization which could indicate a potential event	CCS CSC 14, 16 COBIT 5 DSS05.07 ISA 62443-3-3:2013 SR 6.2 NIST SP 800-53 Rev. 4 AC-2, AU-12, CA-7, CM-3, SC-5, SC-7, SI-4
DE.CM-2: The physical environment is monitored to detect potential cybersecurity events	• Monitoring of physical access controls and the physical environment where DNS components are located is monitored to detect potential incidents • Badge-in access is required for access to critical infrastructure including DNS servers • Access logs are reviewed • Surveillance systems are deployed and reviewed to detect "tailgating," the entry by an unauthorized person before the door closes, for example, and physical removal of assets	ISA 62443-2-1:2009 4.3.3.3.8 NIST SP 800-53 Rev. 4 CA-7, PE-3, PE-6, PE-20

Subcategory	DNS Relevant Activities/Outcomes	Informative References
DE.CM-3: Personnel activity is monitored to detect potential cybersecurity events	• DNS server and management system logs are reviewed periodically to confirm valid user access and activity; any anomalies are investigated as potential security events	ISA 62443-3-3:2013 SR 6.2 ISO/IEC 27001:2013 A.12.4.1 NIST SP 800-53 Rev. 4 AC-2, AU-12, AU-13, CA-7, CM-10, CM-11
DE.CM-4: Malicious code is detected	• Regular virus scans are performed to detect malicious code • Monitor DNS traffic to identify characteristic DNS activity of malware • Block known malware C&C center communications attempts with a DNS firewall	CCS CSC 5 COBIT 5 DSS05.01 ISA 62443-2-1:2009 4.3.4.3.8 ISA 62443-3-3:2013 SR 3.2 ISO/IEC 27001:2013 A.12.2.1 NIST SP 800-53 Rev. 4 SI-3
DE.CM-5: Unauthorized mobile code is detected	• Regular virus scans are performed to detect malicious code • Monitor DNS traffic to identify characteristic DNS activity of malware • Block malware C&C center communications attempts with a DNS firewall	ISA 62443-3-3:2013 SR 2.4 ISO/IEC 27001:2013 A.12.5.1 NIST SP 800-53 Rev. 4 SC-18, SI-4. SC-44
DE.CM-6: External service provider activity is monitored to detect potential cybersecurity events	• External system logs, e.g., for external DNS providers, cloud hosting providers, etc., are reviewed periodically to confirm valid user access and activity; any anomalies are investigated as potential security events	COBIT 5 APO07.06 ISO/IEC 27001:2013 A.14.2.7, A.15.2.1 NIST SP 800-53 Rev. 4 CA-7, PS-7, SA-4, SA-9, SI-4
DE.CM-7: Monitoring for unauthorized personnel, connections, devices, and software is performed	• Monitor for the incidence of unauthorized personnel in secure areas such as datacenters, connections to servers from unauthorized IP addresses, ports, or credentials, any devices not specified within the asset inventory and any software installed on devices beyond that specified in the device baseline • Any such incident of noncompliance should trigger a notification for investigation in accordance with the incident response plan	NIST SP 800-53 Rev. 4 AU-12, CA-7, CM-3, CM-8, PE-3, PE-6, PE-20, SI-4

Subcategory	DNS Relevant Activities/Outcomes	Informative References
DE.CM-8: Vulnerability scans are performed	• Periodic vulnerability scans are performed to detect new vulnerabilities and to verify deployed mitigation controls • Any new vulnerabilities or inadequate mitigation measures are analyzed to assess overall relative risk based on likelihood and business impact and to define new or improved mitigation approaches for development and deployment	COBIT 5 BAI03.10 ISA 62443-2-1:2009 4.2.3.1, 4.2.3.7 ISO/IEC 27001:2013 A.12.6.1 NIST SP 800-53 Rev. 4 RA-5
DETECT: Detection Processes (DE.DP): Detection processes and procedures are maintained and tested to ensure timely and adequate awareness of anomalous events.		
DE.DP-1: Roles and responsibilities for detection are well defined to ensure accountability	• Personnel roles and responsibilities for DNS security event detection within the organization are well defined	CCS CSC 5 COBIT 5 DSS05.01 ISA 62443-2-1:2009 4.4.3.1 ISO/IEC 27001:2013 A.6.1.1 NIST SP 800-53 Rev. 4 CA-2, CA-7, PM-14
DE.DP-2: Detection activities comply with all applicable requirements	• DNS security event detection activities are documented and enforced in accordance with event detection requirements	ISA 62443-2-1:2009 4.4.3.2 ISO/IEC 27001:2013 A.18.1.4 NIST SP 800-53 Rev. 4 CA-2, CA-7, PM-14, SI-4
DE.DP-3: Detection processes are tested	• DNS security event detection activities and systems are tested to characterize detection effectiveness	COBIT 5 APO13.02 ISA 62443-2-1:2009 4.4.3.2 ISA 62443-3-3:2013 SR 3.3 ISO/IEC 27001:2013 A.14.2.8 NIST SP 800-53 Rev. 4 CA-2, CA-7, PE-3, PM-14, SI-3, SI-4

Subcategory	DNS Relevant Activities/Outcomes	Informative References
DE.DP-4: Event detection information is communicated to appropriate parties	• DNS security event detection information is communicated to appropriate parties in accordance with the incident response plan	COBIT 5 APO12.06 ISA 62443-2-1:2009 4.3.4.5.9 ISA 62443-3-3:2013 SR 6.1 ISO/IEC 27001:2013 A.16.1.2 NIST SP 800-53 Rev. 4 AU-6, CA-2, CA-7, RA-5, SI-4
DE.DP-5: Detection processes are continuously improved	• DNS security event detection processes are continuously improved based on technology or process improvements as well as lessons learned from prior events	COBIT 5 APO11.06, DSS04.05 ISA 62443-2-1:2009 4.4.3.4 ISO/IEC 27001:2013 A.16.1.6 NIST SP 800-53 Rev. 4, CA-2, CA-7, PL-2, RA-5, SI-4, PM-14
RESPOND: Response Planning (RS.RP): Response processes and procedures are executed and maintained, to ensure timely response to detected cybersecurity events.		
RS.RP-1: Response plan is executed during or after an event	• As DNS security events are detected and characterized, relevant actions from the incident response plan are executed	COBIT 5 BAI01.10 CCS CSC 18 ISA 62443-2-1:2009 4.3.4.5.1 ISO/IEC 27001:2013 A.16.1.5 NIST SP 800-53 Rev. 4 CP-2, CP-10, IR-4, IR-8
RESPOND: Communications (RS.CO): Response activities are coordinated with internal and external stakeholders, as appropriate, to include external support from law enforcement agencies.		
RS.CO-1: Personnel know their roles and order of operations when a response is needed	• Personnel roles and responsibilities for DNS security event response within the organization are well defined within the incident response plan	ISA 62443-2-1:2009 4.3.4.5.2, 4.3.4.5.3, 4.3.4.5.4 ISO/IEC 27001:2013 A.6.1.1, A.16.1.1 NIST SP 800-53 Rev. 4 CP-2, CP-3, IR-3, IR-8
RS.CO-2: Events are reported consistent with established criteria	• DNS security events are detected and characterized in accordance with established criteria defined in the incident response plan	ISA 62443-2-1:2009 4.3.4.5.5 ISO/IEC 27001:2013 A.6.1.3, A.16.1.2 NIST SP 800-53 Rev. 4 AU-6, IR-6, IR-8

Subcategory	DNS Relevant Activities/Outcomes	Informative References
RS.CO-3: Information is shared consistent with response plans	• DNS security event response information is communicated to appropriate parties in accordance with the incident response plan	ISA 62443-2-1:2009 4.3.4.5.2 ISO/IEC 27001:2013 A.16.1.2 NIST SP 800-53 Rev. 4 CA-2, CA-7, CP-2, IR-4, IR-8, PE-6, RA-5, SI-4
RS.CO-4: Coordination with stakeholders occurs consistent with response plans	• DNS security event response information is communicated to appropriate stakeholders in accordance with the incident response plan	ISA 62443-2-1:2009 4.3.4.5.5 NIST SP 800-53 Rev. 4 CP-2, IR-4, IR-8
RS.CO-5: Voluntary information sharing occurs with external stakeholders to achieve broader cybersecurity situational awareness	• DNS security event response information is communicated to external stakeholders in accordance with the incident response plan to facilitate industry awareness of the attack and effective defensive measures	NIST SP 800-53 Rev. 4 PM-15, SI-5
RESPOND: Analysis (RS.AN): Analysis is conducted to ensure adequate response and support recovery activities.		
RS.AN-1: Notifications from detection systems are investigated	• DNS event detection systems are investigated to characterize the event as a potential attack or threat	COBIT 5 DSS02.07 ISA 62443-2-1:2009 4.3.4.5.6, 4.3.4.5.7, 4.3.4.5.8 ISA 62443-3-3:2013 SR 6.1 ISO/IEC 27001:2013 A.12.4.1, A.12.4.3, A.16.1.5 NIST SP 800-53 Rev. 4 AU-6, CA-7, IR-4, IR-5, PE-6, SI-4

Subcategory	DNS Relevant Activities/Outcomes	Informative References
RS.AN-2: The impact of the incident is understood	• Upon incident detection, the incident is analyzed to assess and understand the impact of the incident • The incident response plan is followed and impacted groups involved in responding to contain, eradicate, and recover from the incident, while communicating status in accordance with the response plan • New information or lessons learned are incorporated into an update of the response plan based on review, concurrence, and approval by appropriate members of the organization	ISA 62443-2-1:2009 4.3.4.5.6, 4.3.4.5.7, 4.3.4.5.8 ISO/IEC 27001:2013 A.16.1.6 NIST SP 800-53 Rev. 4 CP-2, IR-4
RS.AN-3: Forensics are performed	• Forensic analysis on detected incidents is performed to go beyond the symptoms of the incident to identify the ultimate cause and to enumerate those vulnerabilities exploited or attacked. This analysis is useful for identifying new or morphed attack vectors and vulnerabilities, and to qualify the effectiveness of any defensive controls that were intended to protect against such an attack	ISA 62443-3-3:2013 SR 2.8, SR 2.9, SR 2.10, SR 2.11, SR 2.12, SR 3.9, SR 6.1 ISO/IEC 27001:2013 A.16.1.7 NIST SP 800-53 Rev. 4 AU-7, IR-4
RS.AN-4: Incidents are categorized consistent with response plans	• Incidents are categorized in a manner consistent with incident response plans. This is helpful in terms of prioritizing actions and inclusion of appropriate staff to analyze, contain, eradicate, and resolve the incident in a timely manner	ISA 62443-2-1:2009 4.3.4.5.6 ISO/IEC 27001:2013 A.16.1.4 NIST SP 800-53 Rev. 4 CP-2, IR-4, IR-5, IR-8

Subcategory	DNS Relevant Activities/Outcomes	Informative References
RESPOND: Mitigation (RS.MI): Activities are performed to prevent expansion of an event, mitigate its effects, and eradicate the incident.		
RS.MI-1: Incidents are contained	• Deployment of DNS components in accordance with defined trust sectors provides containment to the corresponding trust sector • Further containment steps must be undertaken based on the incident itself to prevent broader impact on multiple DNS servers, resolvers, or other network systems	ISA 62443-2-1:2009 4.3.4.5.6 ISA 62443-3-3:2013 SR 5.1, SR 5.2, SR 5.4 ISO/IEC 27001:2013 A.16.1.5 NIST SP 800-53 Rev. 4 IR-4
RS.MI-2: Incidents are mitigated	• As the incident is contained, contingency plans implemented, and forensic analyses conducted, mitigation approaches for the vulnerability that led to the successful incident are defined, evaluated, agreed upon, and implemented • Based on the particular incident, mitigate in accordance with recommended mitigation tactics • The vulnerability list, risk assessment, incident response plans are updated accordingly	ISA 62443-2-1:2009 4.3.4.5.6, 4.3.4.5.10 ISO/IEC 27001:2013 A.12.2.1, A.16.1.5 NIST SP 800-53 Rev. 4 IR-4
RS.MI-3: Newly identified vulnerabilities are mitigated or documented as accepted risks	• Newly identified vulnerabilities are incorporated into the known vulnerability list • Each new vulnerability is analyzed with respect to likelihood and business impact to define relative risk. Based on this assessed risk, the vulnerability is proactively mitigated or documented as an accepted risk	ISO/IEC 27001:2013 A.12.6.1 NIST SP 800-53 Rev. 4 CA-7, RA-3, RA-5
RESPOND: Improvements (RS.IM): Organizational response activities are improved by incorporating lessons learned from current and previous detection/response activities.		
RS.IM-1: Response plans incorporate lessons learned	• After incident recovery, a postmortem discussion with involved staff is conducted to review the incident, define possible defensive and mitigation steps to improve response and recommended response plan updates to incorporate lessons learned	COBIT 5 BAI01.13 ISA 62443-2-1:2009 4.3.4.5.10, 4.4.3.4 ISO/IEC 27001:2013 A.16.1.6 NIST SP 800-53 Rev. 4 CP-2, IR-4, IR-8

Subcategory	DNS Relevant Activities/Outcomes	Informative References
RS.IM-2: Response strategies are updated	• Incident response strategies are reviewed and updated as appropriate	NIST SP 800-53 Rev. 4 CP-2, IR-4, IR-8
RECOVER: Recovery Planning (RC.RP): Recovery processes and procedures are executed and maintained to ensure timely restoration of systems or assets affected by cybersecurity events.		
RC.RP-1: Recovery plan is executed during or after an event	• The incident recovery plan is executed during or after an event • During the event, contingencies and workarounds are put in place to restore service levels in the face of a disruption, compromise, or outage • After incident eradication, affected systems are restored to prior function to fully recovery to a known working state	CCS CSC 8 COBIT 5 DSS02.05, DSS03.04 ISO/IEC 27001:2013 A.16.1.5 NIST SP 800-53 Rev. 4 CP-10, IR-4, IR-8
RECOVER: Improvements (RC.IM): Recovery planning and processes are improved by incorporating lessons learned into future activities.		
RC.IM-1: Recovery plans incorporate lessons learned	• Recovery plans are updated to incorporate lessons learned	COBIT 5 BAI05.07 ISA 62443-2-1 4.4.3.4 NIST SP 800-53 Rev. 4 CP-2, IR-4, IR-8
RC.IM-2: Recovery strategies are updated	• Recovery strategies are reviewed and updated should any improvements be borne out of the analysis of the incident recovery	COBIT 5 BAI07.08 NIST SP 800-53 Rev. 4 CP-2, IR-4, IR-8
RECOVER: Communications (RC.CO): Restoration activities are coordinated with internal and external parties, such as coordinating centers, Internet service providers, owners of attacking systems, victims, other CSIRTs, and vendors.		
RC.CO-1: Public relations are managed	• Communications to customers and to the public in general are carefully managed to convey information regarding the incident, status of response and recovery as well as planned actions	COBIT 5 EDM03.02
RC.CO-2: Reputation after an event is repaired	• Typically, the provision of meaningful information regarding the incident and what has been done to recover from the incident helps with preserving reputation but other steps may be necessary	COBIT 5 MEA03.02

Subcategory	DNS Relevant Activities/Outcomes	Informative References
RC.CO-3: Recovery activities are communicated to internal stakeholders and executive and management teams	• Communications to internal stakeholders including executives and management are open and direct regarding the incident, status of response, and recovery and planned actions including evaluation of alternative approaches if the attack persists	NIST SP 800-53 Rev. 4 CP-2, IR-4

B

DNS RESOURCE RECORD TYPES

Table B.1 summarizes the currently defined set of resource records in alphabetical order by resource record type (RRType – also corresponds to valid QType when a querier seeks this type of information from DNS, that is, within the Question section of a DNS message). While not all resource records are IETF standards or even defined within the IETF, those that have been assigned an RR Type ID number by IANA are listed here. Current IETF status is provided along with the defining document which can be accessed for more details.

TABLE B.1   Resource Record and Query Type Summary

RRType (or QType)	RR Purpose (i.e., RData Contents)	RR Type ID	IETF Status	Defining Document
A	IPv4 address for a given hostname	1	Standard	RFC 1035 (17)
AAAA	IPv6 address for a given hostname	28	Draft standard	RFC 3596 (97)
A6	IPv6 address or portion thereof for iterative IPv6 address resolution for a given hostname	38	Obsolete	RFC 2874 (98), 6563 (99)
AFSDB	Server hostname for a given AFS and DCE domain	18	Experimental	RFC 1183 (100)

(continued)

DNS Security Management, First Edition. Michael Dooley and Timothy Rooney.
© 2017 by The Institute of Electrical and Electronic Engineers, Inc. Published 2017 by John Wiley & Sons, Inc.

TABLE B.2   (*Continued*)

RRType (or QType)	RR Purpose (i.e., RData Contents)	RR Type ID	IETF Status	Defining Document
APL	Address prefix lists for a given domain	42	Experimental	RFC 3123 (101)
ATMA	Asynchronous transfer mode (ATM) address for a host	34	Not submitted	ATM Name System Specification by the ATM Forum (102)
AVC	Application visibility and control	258	Not submitted	www.dns-as.org
CAA	Certification authority restriction	257	Proposed standard	RFC 6844 (103)
CDNSKEY	DNSKEY(s) the child wants reflected in DS	60	Informational	RFC 7344 (59)
CDS	Child delegation signer (DS)	59	Informational	RFC 7344 (59)
CERT	Certificate or certificate revocation list	37	Standards track	RFC 4398 (92)
CNAME	Alias host name for a host	5	Standard	RFC 1035 (17)
CSYNC	Child-to-parent synchronization	62	Proposed standard	RFC 7477 (104)
DHCID	Associates a DHCP client's identity with a DNS name	49	Standards track	RFC 4701 (105)
DLV	Authoritative zone signature for a trust anchor	32769	Informational (DNSSEC)	RFC 4431 (106)
DNAME	Alias domain name	39	Proposed standard	RFC 6672 (19)
DNSKEY	Authoritative zone signature within a chain of trust	48	Standards track (DNSSEC)	RFC 4034 (107)
DS	Signature for delegated child zone	43	Standards track (DNSSEC)	RFC 4034 (107)
EID	Endpoint identifier	31	Internet draft (expired)	draft-ietf-nimrod-dns-01.txt
EUI48	EUI-48 layer 2 addresses	108	Informational	RFC 7043 (108)
EUI64	EUI-64 layer 2 addresses	109	Informational	RFC 7043 (108)
GID	Group identifier	102	RESERVED	IANA-reserved
GPOS	Latitude/Longitude/Altitude for a given host – superseded by LOC	27	Experimental	RFC 1712 (109)
HINFO	CPU and OS information for a host	13	Standard	RFC 1035 (17)
HIP	Host identity protocol	55	Experimental	RFC 8005 (110)

(*continued*)

TABLE B.3 (Continued)

RRType (or QType)	RR Purpose (i.e., RData Contents)	RR Type ID	IETF Status	Defining Document
IPSECKEY	Public key for a given DNS name for use with IPSec	45	Proposed standard	RFC 4025 (93)
ISDN	Integrated services digital network (ISDN) address and subaddress for a given host	20	Experimental	RFC 1183 (100)
KEY	Superseded by DNSKEY within DNSSEC but still used by SIG(0) and TKEY	25	Proposed standard	RFC 2536 (111)
KX	Intermediary domain to obtain a key for a host in given domain	36	Informational	RFC 2230 (112)
L32	Identifier-locator network protocol	105	Experimental	RFC 6742 (113)
L64	Identifier-locator network protocol	106	Experimental	RFC 6742 (113)
LOC	Latitude/Longitude/Altitude and precision for a given host	29	Uncommon	RFC 1876 (114)
LP	Identifier-locator network protocol	107	Experimental	RFC 6742 (113)
MB	Mailbox name for a given email ID	7	Experimental	RFC 1035 (17)
MD	Mail delivery host for a given domain	3	Obsolete	RFC 1035 (17)
MF	Host that will accept mail for forwarding to a given domain	4	Obsolete	RFC 1035 (17)
MG	Mail group mailbox name for a given email ID	8	Experimental	RFC 1035 (17)
MINFO	Mailbox names for sending account requests or error reports for a given mailbox name	14	Experimental	RFC 1035 (17)
MR	Alias for a mailbox name	9	Experimental	RFC 1035 (17)
MX	Mail exchange for email host resolution	15	Standard	RFC 1035 (17)
NAPTR	Uniform resource identifier for a generic string – used for DDDS, ENUM applications	35	Standards track	RFC 3403 (115)
NID	Identifier-locator network protocol	104	Experimental	RFC 6742 (113)
NIMLOC	Nimrod routing architecture	32	Internet draft (expired)	draft-ietf-nimrod-dns-01.txt
NINFO	DNS zone status	56	Not submitted	None
NS	Name server for a given domain name	2	Standard	RFC 1035 (17)

(continued)

TABLE B.4   *(Continued)*

RRType (or QType)	RR Purpose (i.e., RData Contents)	RR Type ID	IETF Status	Defining Document
NSAP	Network services access point address for a host	22	Uncommon	RFC 1706 (116)
NSAP-PTR	Hostname for a given NSAP address	23	Uncommon	RFC 1706 (116)
NSEC	Authenticated confirmation or denial of existence of a resource record set for DNSSEC	47	Standards track (DNSSEC)	RFC 4034 (107)
NSEC3	Authenticated denial of existence of a resource record set for DNSSEC (without trivial zone enumeration obtainable with NSEC)	50	Standards track (DNSSEC)	RFC 5155 (117)
NSEC3 PARAM	NSEC3 parameters used to calculate hashed owner names	51	Standards track (DNSSEC)	RFC 5155 (117)
NULL	Up to 65535 bytes of anything for a given host	10	Experimental	RFC 1035 (17)
NXT	Superseded by NSEC	30	Obsolete (DNSSEC)	RFC 3755 (118)
OPENPGPKEY	OpenPGP public key for email addresses (DANE)	61	Experimental	RFC 7929 (119)
PTR	Hostname for a given IPv4 or IPv6 address	12	Standard	RFC 1035 (17)
PX	X.400 mapping for a given domain name	26	Uncommon	RFC 2163 (120)
RKEY	Public key for NAPTR resource record encryption	57	Internet draft (expired)	draft-reid-dnsext-rkey-00.txt
RP	Email address and TXT record pointer for more info for a host	17	Experimental	RFC 1183 (100)
RRSIG	Signature for a resource record set of a given domain name, class, and RR type	46	Standards track (DNSSEC)	RFC 4034 (107)
RT	Proxy hostname for a given host that is not always connected	21	Experimental	RFC 1183 (100)
SIG	Superseded by RRSIG within DNSSEC; used by SIG(0) and TKEY	24	Proposed standard	RFC 2536 (111)
SINK	Miscellaneous structured information	40	Internet draft (expired)	https://tools.ietf.org/html/draft-eastlake-kitchen-sink-02

(continued)

TABLE B.5 (Continued)

RRType (or QType)	RR Purpose (i.e., RData Contents)	RR Type ID	IETF Status	Defining Document
SIMIMEA	Certificate association with domain names for S/MIME	53	Internet draft	https://tools.ietf. org/html/draft-ietf- dane-smime-16
SOA	Authority information for a zone	6	Standard	RFC 1035 (17)
SPF	Sender policy framework – enables a domain owner to identify hosts authorized to send emails from the domain	99	Experimental	RFC 7208 (89)
SRV	Host providing specified services in a domain	33	Standards track	RFC 2782 (121)
SSHFP	Secure shell fingerprints – enables verification of SSH host keys using DNSSEC	44	Standards track	RFC 4255 (122)
TA	DNSSEC trust authorities	32768	Internet draft (expired)	http://www.watson. org/~weiler/ INI1999-19.pdf
TALINK	Trust anchor history link	58	Internet draft (expired)	https://tools.ietf. org/html/draft- wijngaards-dnsop- trust-history-00
TLSA	Transport layer security association (DANE)	52	Proposed standard	RFC 6698 (80)
TXT	Arbitrary text associated with a host	16	Standard	RFC 1035 (17)
UID	User ID	101	RESERVED	IANA-Reserved
UINFO	User info	100	RESERVED	IANA-Reserved
UNSPEC	Unspecified	103	RESERVED	IANA-Reserved
URI	Uniform resource identifier	256	Informational	RFC 7553 (123)
WKS	Services available via a given protocol at a specified IP address for a host – SRV RR more commonly used today	11	Standard	RFC 1035 (17)
X25	X.25 packet switched data network (PSDN)	19	Experimental	RFC 1183 (100)

BIBLIOGRAPHY

1. National Institute of Standards and Technology. *Framework for Improving Critical Infrastructure Cybersecurity*. February 12, 2014.

2. ISACA. *COBIT 5: An ISACA Framework*. COBIT Online. [Online] [Cited: August 6, 2016]. https://cobitonline.isaca.org/

3. ANSI/ISA. *Security for Industrial Automation and Control Systems: System Security Requirements and Security Levels*. 2013. ANSI/ISA-62443-3 (99.03.03)-2013.

4. ISO/IEC. *Information Technology – Security Techniques – Information Security Management Systems – Requirements*. April 2013. ISO/IEC 27001.

5. National Institute of Standards and Technology. *Security and Privacy Controls for Federal Information Systems and Organizations*. 2014. NIST Special Publication 800-53 Revision 4.

6. Computer Security Division, Information Technology Laboratory, National Institute of Standards and Technology. *Standards for Security Categorization of Federal Information and Information Systems*. Federal Information Processing Standards Publication, February 2004. FIPS Pub. 199.

7. Chandramouli, R., and Rose, S. *Secure Domain Name System (DNS) Deployment Guide*. Gaithersburg, MD: National Institute of Standards and Technology, September 2013. NIST Special Publication 800-81-2.

8. Rooney, T. *IP Address Management Principles and Practice*. Wiley & Sons/IEEE Press, 2011.

9. Klensin, J. *Internationalized Domain Names for Applications (IDNA): Definitions and Document Framework*. IETF, August 2010. RFC 5890.

10. Klensin, J. *Internationalized Domain Names in Applications (IDNA): Protocol*. IETF, August 2010. RFC 5891.

11. Faltstrom, P., Ed. *The Unicode Code Points and Internationalized Domain Names for Applications (IDNA)*. IETF, August 2010. RFC 5892.

12. Karp, C. *Right-to-Left Scripts for Internationalized Domain Names for Applications (IDNA)*. Edited by H. Alvestrand. IETF, August 2010. RFC 5893.

13. Klensin, J. *Internationalized Domain Names for Applications (IDNA): Background, Explanation, and Rationale*. IETF, August 2010. RFC 5894.

DNS Security Management, First Edition. Michael Dooley and Timothy Rooney.
© 2017 by The Institute of Electrical and Electronic Engineers, Inc. Published 2017 by John Wiley & Sons, Inc.

14. Resnick, P., and Hoffman, P. *Mapping Characters for Internationalized Domain Names in Applications (IDNA) 2008*. IETF, September 2010. RFC 5895.

15. Costello, A. *Punycode: A Bootstring Encoding of Unicode for Internationalized Domain Names in Applications (IDNA)*. IETF, March 2003. RFC 3492.

16. International Telecommunications Union. *Internationalized Domain Names (IDN)*. [Online] [Cited: October 21, 2009]. http://www.itu.int/ITU-T/special-projects/idn/introduction.html

17. Mockapetris, P. *Domain Names – Implementation and Specification*. IETF, November 1987. RFC 1035.

18. Thomson, S., Rekhter, Y., and Bound, J. *Dynamic Updates in the Domain Name System (DNS UPDATE)*. Edited by P. Vixie. IETF, April 1997. RFC 2136.

19. Rose, S., and Wijngaards, W. *DNAME Redirection in the DNS*. IETF, June 2012. RFC 6672.

20. Damas, J., Graff, M., and Vixie, P. *Extension Mechanisms for DNS (EDNS(0))*. IETF, April 2013. RFC 6891.

21. Vixie, P., Gudmundsson, O., Eastlake 3rd, D., and Wellington, B. *Secret Key Transaction Authentication for DNS (TSIG)*. IETF, May 2000. RFC 2845.

22. Eastlake 3rd, D. *Secret Key Establishment for DNS (TKEY RR)*. IETF, September 2000. RFC 2930.

23. Eastlake 3rd, D. *HMAC SHA TSIG Algorithm Identifiers*. IETF, August 2006. RFC 4635.

24. Eastlake 3rd, D., and Andrews, M. *Domain Name System (DNS) Cookies*. IETF, May 2016. RFC 7873.

25. Eastlake 3rd, D. *Domain Name System (DNS) IANA Considerations*. IETF, April 2013. RFC 6895.

26. Ohta, M. *Incremental Zone Transfer in DNS*. IETF, August 1996. RFC 1995.

27. Austein, R. *DNS Name Server Identifier (NSID) Option*. IETF, August 2007. RFC 5001.

28. Crocker, S., and Rose, S. *Signaling Cryptographic Algorithm Understanding in DNS Security Extensions (DNSSEC)*. IETF, July 2013. RFC 6975.

29. Contavalli, C., van der Gaast, W., Lawrence, D., and Kumari, W. *Client Subnet in DNS Queries*. IETF, May 2016. RFC 7871.

30. Andrews, M. *Extension Mechanisms for DNS (EDNS) EXPIRE Option*. IETF, July 2014. RFC 7314.

31. Wouters, P., Abley, J., Dickinson, S., and Bellis, R. *The edns-tcp-keepalive EDNS0 Option*. IETF, April 2016. RFC 7828.

32. Mayrhofer, A. *The EDNS(0) Padding Option*. IETF, May 2016. RFC 7830.

33. Wouters, P. *CHAIN Query Requests in DNS*. IETF, June 2016. RFC 7901.

34. Bortzmeyer, S. *DNS Query Name Minimisation to Improve Privacy*. IETF, March 2016. RFC 7816.

35. York, K. *Dyn Statement on 10/21/2016 DDoS Attack*. Dyn. [Online] [Cited: October 23, 2016]. https://dyn.com/blog/dyn-statement-on-10212016-ddos-attack/

36. Drozhzhin, A. *Switcher Hacks Wi-Fi Routers, Switches DNS*. Kaspersky Lab. [Online] [Cited: December 29, 2016]. https://blog.kaspersky.com/switcher-trojan-attacks-routers/13771/

37. Open Resolver Project. [Online]. http://openresolverproject.org/

38. United States Computer Emergency Readiness Team (US-CERT). *DNS Amplification Attacks*. [Online] [Cited: September 16, 2016]. https://www.us-cert.gov/ncas/alerts/TA13-088A

39. Ferguson, P., and Senie, D. *Network Ingress Filtering: Defeating Denial of Service Attacks Which Employ IP Source Address Spoofing*. IETF, May 2000. RFC 2827, BCP 38.

40. Internet Corporation for Assigned Names and Numbers (ICANN). *Factsheet: Root Server Attack on 6 February 2007*. [Online] [Cited: June 26, 2016]. https://www.icann.org/en/system/files/files/factsheet-dns-attack-08mar07-en.pdf

41. United States Computer Emergency Readiness Team (US-CERT). *Mailing Lists and Feeds*. [Online] [Cited: September 7, 2016]. https://www.us-cert.gov/mailing-lists-and-feeds

42. United States Computer Emergency Readiness Team (US-CERT), United States Department of Homeland Security. [Online] [Cited: October 22, 2016]. https://www.us-cert.gov/

43. MITRE Corporation. *Common Vulnerabilities and Exposures*. [Online] [Cited: October 22, 2016]. https://cve.mitre.org/

44. Boran, S. *Running BIND9 DNS Server Securely*. Boran Consulting. [Online] [Cited: September 9, 2016]. http://www.boran.com/security/sp/bind9_20010430.html

45. Scholten, M. *Pdns-dev Email Archive*. [Online] [Cited: November 12, 2016]. https://mailman.powerdns.com/pipermail/pdns-dev/2012-June/001179.html

46. Zalewski, M. *Strange Attractors and TCP/IP Sequence Number Analysis*. [Online] [Cited: October 30, 2016]. http://lcamtuf.coredump.cx/oldtcp/tcpseq.html

47. United States Computer Emergency Readiness Team. *Various DNS Service Implementations Generate Multiple Simultaneous Queries for the Same Resource Record*. Vulnerability Note VU#457875. CERT Vulnerability Notes Database. [Online] [Cited: October 30, 2016]. http://www.kb.cert.org/vuls/id/457875

48. Kaminsky, D. *Black Ops 2008: It's the End of the Cache as We Know It*. LinkedIn SlideShare. [Online] [Cited: August 13, 2016]. http://www.slideshare.net/dakami/dmk-bo2-k8

49. National Institute of Standards and Technology. *Digital Signature Standard (DSS)*. Gaithersburg: National Institute of Standards and Technology, July 2013. FIPS Pub. 186-4.

50. StJohns, M. *Automated Updates of DNS Security (DNSSEC) Trust Anchors*. IETF, September 2007. RFC 5011.

51. Internet Systems Consortium. *BIND DNSSEC Guide*. Redwood City: Internet Systems Consortium, 2015.

52. NLnet Labs. *Unbound-Anchor Man Page*. Amsterdam: NLnet Labs, 2016.

53. NLnet Labs. *Unbound: How to Enable DNSSEC*. Unbound Documentation. [Online] [Cited: October 2, 2016]. http://unbound.net/documentation/howto_anchor.html

54. NIST Advanced Network Technologies Division. *Estimating IPv6 & DNSSEC Deployment*. NIST Advanced Network Technologies Division, Information Technology

Laboratory. [Online] [Cited: October 2, 2016]. https://usgv6-deploymon.antd.nist.gov/snap-all.html

55. APNIC Labs. *Use of DNSSEC Validation for World.* [Online] [Cited: October 2, 2016]. http://stats.labs.apnic.net/dnssec/XA?c=XA&x=1&g=1&r=1&w=7&g=0

56. Internet Assigned Numbers Authority. *Root Zone Database.* [Online] [Cited: October 3, 2016]. http://www.iana.org/domains/root/db

57. ICANN. *TLD DNSSEC Report.* ICANN Research. [Online] [Cited: December 18, 2016]. http://stats.research.icann.org/dns/tld_report/

58. Kolkman, O., Mekking, W., and Gieben, R. *DNSSEC Operational Practices, Version 2.* IETF, December 2012. RFC 6781.

59. Kumari, W., Gudmundsson, O., and Barwood, G. *Automating DNSSEC Delegation Trust Maintenance.* IETF, September 2014. RFC 7344.

60. Dell EMC. *PKCS #11: Cryptographic Token Interface Standard.* RSA Laboratories. [Online] [Cited: November 5, 2016]. https://www.emc.com/emc-plus/rsa-labs/standards-initiatives/pkcs-11-cryptographic-token-interface-standard.htm

61. Internet Systems Consortium. *dnssec-keymgr Manual Pages.* [Online] [Cited: November 5, 2016]. https://ftp.isc.org/isc/bind9/cur/9.11/doc/arm/man.dnssec-keymgr.html

62. PowerDNS. *pdnsutil – PowerDNS dnssec Command and Control.* [Online] [Cited: November 5, 2016]. https://doc.powerdns.com/md/manpages/pdnsutil.1/

63. CZ NIC. *Knot DNS Configuration.* Knot DNS 2.3.2 Documentation. [Online] [Cited: November 5, 2016]. https://www.knot-dns.cz/docs/2.x/html/configuration.html

64. Farnham, G. *Detecting DNS Tunneling.* SANS Institute, February 2013.

65. Jaworski, S. *Using Splunk to Detect DNS Tunneling.* SANS Institute, May 2016.

66. Krebs, B. *KrebsOnSecurity Hit with Record DDoS.* Krebs on Security. [Online] [Cited: October 15, 2016]. https://krebsonsecurity.com/2016/09/krebsonsecurity-hit-with-record-ddos/

67. United States Computer Emergency Readiness Team. *Backoff Point-of-Sale Malware.* US-CERT. Original release date: July 2014. Last revised: September 2016. TA14-212A.

68. United States Computer Emergency Readiness Team. *Crypto Ransomware.* US-CERT. Original release date: October 2014. Last revised: September 2016. TA14-295A.

69. United States Computer Emergency Readiness Team. *Dridex P2P Malware.* US-CERT. Original release date: October 2015. Last revised: September 2016. TA15-286A.

70. United States Computer Emergency Readiness Team. *Ransomware and Recent Variants.* US-CERT. Original release date: March 2016. Last revised: September 2016. TA16-091A.

71. United States Computer Emergency Readiness Team. *Apple iOS "Masque Attack" Technique.* US-CERT. Original release date: November 2014. Last revised: September 2016. TA14-317A.

72. United States Computer Emergency Readiness Team. *Heightened DDoS Threat Posed by Mirai and Other Botnets.* US-CERT. Original release date: October 2016. Last revised: November 2016. TA16-288A.

73. Microsoft Corporation. *TrojanSpy: Win32/Nivdort.A.* Malware Protection Center. [Online] [Cited: October 30, 2016]. https://www.microsoft.com/security/portal/threat/encyclopedia/entry.aspx?Name=TrojanSpy%3aWin32%2fNivdort.A

74. United States Computer Emergency Readiness Team. *Simda Botnet*. US-CERT. Original release date: April 2015. Last revised: September 2016. TA15-105A.

75. Weimer, F. *Passive DNS Replication*. April 2005.

76. Perdisci, R., Corona, I., and Giacinto, G. *Early Detection of Malicious Flux Networks via Large-Scale Passive DNS Traffic Analysis*. IEEE Transactions on Dependable and Secure Computing (IEEE-TDSC), September–October 2012, pp. 714–726.

77. Almomani, A. *Fast-Flux Hunter: A System for Filtering Online Fast-Flux Botnet*. Neural Computing and Applications, August 2016. DOI: 10.1007/s00521-016-2531-1.

78. Soltanaghaei, E., and Kharrazi, M. *Detection of Fast-Flux Botnets Through DNS Traffic Analysis*. Scientia Iranica, January 2016.

79. Antonakakis, M., Perdisci, R., Nadji, Y., Vasiloglou, N., Abu-Nimeh, S., Lee, W., and Dagon, D. *From Throw-Away Traffic to Bots: Detecting the Rise of DGA-Based Malware*. 2012.

80. Hoffman, P., and Schlyter, J. *The DNS-Based Authentication of Named Entities (DANE) Transport Layer Security (TLS) Protocol: TLSA*. IETF, August 2012. RFC 6698.

81. Dukhovni, V., and Hardaker, W. *The DNS-Based Authentication of Named Entities (DANE) Protocol: Updates and Operational Guidance*. IETF, October 2015. RFC 7671.

82. Bright, P. *Independent Iranian Hacker Claims Responsibility for Comodo Hack*. ARS Technica. [Online] [Cited: October 13, 2016]. http://arstechnica.com/security/2011/03/independent-iranian-hacker-claims-responsibility-for-comodo-hack/

83. Bright, P. *Another Fraudulent Certificate Raises the Same Old Questions About Certificate Authorities*. ARS Technica. [Online] [Cited: October 13, 2016]. http://arstechnica.com/security/2011/08/earlier-this-year-an-iranian/

84. Leydon, J. *Symantec Fires Staff Caught Up in Rogue Google SSL Cert Snafu*. The Register. [Online] [Cited: October 13, 2016]. http://www.theregister.co.uk/2015/09/21/symantec_fires_workers_over_rogue_certs/

85. IETF. *DNS-Based Authentication of Named Entities*. IETF Datatracker. [Online] [Cited: October 15, 2016]. https://datatracker.ietf.org/wg/dane/documents/

86. Hoffman, P., and Schlyter, J. *Using Secure DNS to Associate Certificates with Domain Names for S/MIME*. IETF, July 2016. draft-ietf-dane-smime-12.

87. Rose, S., Barker, W. C., Jha, S., Irrechukwu, C., and Waltermire, K. *Domain Name Systems-Based Electronic Mail Security*. NCCoE, NIST, November 2016. NIST Special Publication 1800-6 (Draft).

88. Lewis, C., and Sergeant, M. *Overview of Best Email DNS-Based List (DNSBL) Operational Practices*. IETF, January 2012. RFC 6471.

89. Kitterman, S. *Sender Policy Framework (SPF) for Authorizing Use of Domains in Email, Version 1*. April 2014. RFC 7208.

90. Allman, E., Fenton, J., Delany, M., and Levine, J. *DomainKeys Identified Mail (DKIM) Author Domain Signing Practices (ADSP)*. IETF, August 2009. RFC 5617.

91. Kucherawy, M., and Zwicky, E. *Domain-Based Message Authentication, Reporting, and Conformance (DMARC)*. IETF, March 2015. RFC 7489.

92. Josefsson, S. *Storing Certificates in the Domain Name System (DNS)*. IETF, March 2006. RFC 2538.

93. Richardson, M. *A Method for Storing IPsec Keying Material in DNS*. IETF, February 2005. RFC 4025.

94. Eastlake 3rd, D. *DNS Request and Transaction Signatures (SIG(0)s)*. IETF, September 2000. RFC 2931.

95. Splunk. [Online]. www.splunk.com

96. Stratos IP. [Online]. www.stratosip.com

97. Thomson, S., Huitema, C., Ksinant, V., and Souissi, M. *DNS Extensions to Support IP Version 6*. IETF, October 2003. RFC 3596.

98. Crawford, M., and Huitema, C. *DNS Extensions to Support IPv6 Address Aggregation and Renumbering*. IETF, July 2000. RFC 2874.

99. Jiang, S., Conrad, D., and Carpenter, B. *Moving A6 to Historic Status*. IETF, March 2012. RFC 6563.

100. Everhart, C., Mamakos, L., and Ullmann, R. *New DNS RR Definitions*. Edited by P. Mockapetris. IETF, October 1990. RFC 1183.

101. Koch, P. *A DNS RR Type for Lists of Address Prefixes (APL RR)*. IETF, June 2001. RFC 3123.

102. The ATM Forum. *ATM Name System V2.0*. July 2000. AF-DANS-0152.000.

103. Hallman-Baker, P., and Stradling, R. *DNS Certification Authority Authorization (CAA) Resource Record*. IETF, July 2013. RFC 6844.

104. Hardaker, W. *Child-to-Parent Synchronization in DNS*. IETF, March 2015. RFC 7477.

105. Stapp, M., Lemon, T., and Gustafsson, A. *A DNS Resource Record (RR) for Encoding Dynamic Host Configuration Protocol (DHCP) Information (DHCID RR)*. IETF, October 2006. RFC 4701.

106. Andrews, M., and Weiler, S. *The DNSSEC Lookaside Validation (DLV) DNS Resource Record*. IETF, February 2006. RFC 4431.

107. Arends, R., Austein, R., Larson, M., Massey, D., and Rose, S. *Resource Records for the DNS Security Extensions*. IETF, March 2005. RFC 4034.

108. Abley, J. *Resource Records for EUI-48 and EUI-64 Addresses in the DNS*. IETF, October 2013. RFC 7043.

109. Farrell, C., Schulze, M., Pleitner, S., and Baldoni, D. *DNS Encoding of Geographical Location*. IETF, November 1994. RFC 1712.

110. Laganier, J. *Host Identity Protocol (HIP) Domain Name System (DNS) Extension*. IETF, October 2016. RFC 8005.

111. Eastlake 3rd, D. *DSA KEYs and SIGs in the Domain Name System (DNS)*. IETF, March 1999. RFC 2536.

112. Atkinson, R. *Key Exchange Delegation Record for the DNS*. IETF, November 1997. RFC 2230.

113. Atkinson, R. J., Bhatti, S. N., and Rose, S. *DNS Resource Records for the Identifier-Locator Network Protocol (ILNP)*. IETF, November 2012. RFC 6742.

114. Davis, C., Vixie, P., Goodwin, T., and Dickinson, I. *A Means for Expressing Location Information in the Domain Name System*. IETF, January 1996. RFC 1876.

115. Mealling, M. *Dynamic Delegation Discovery System (DDDS) Part Three: The Domain Name System (DNS) Database.* IETF, October 2002. RFC 3403.

116. Manning, B., and Colella, R. *DNS NSAP Resource Records.* IETF, October 1994. RFC 1706.

117. Laurie, B., Sisson, G., Arends, R., and Blacka, D. *DNS Security (DNSSEC) Hashed Authenticated Denial of Existence.* IETF, March 2008. RFC 5155.

118. Weiler, S. *Legacy Resolver Compatibility for Delegation Signer (DS).* IETF, May 2004. RFC 3755.

119. Wouters, P. *DNS-Based Authentication of Named Entities (DANE) Bindings for OpenPGP.* IETF, April 2016. RFC 7929.

120. Allocchio, C. *Using the Internet DNS to Distribute MIXER Conformant Global Address Mapping (MCGAM).* IETF, January 1998. RFC 2163.

121. Gulbrandsen, A., Vixie, P., and Esibov, L. *A DNS RR for Specifying the Location of Services (DNS SRV).* IETF, February 2000. RFV 2782.

122. Schlyter, J., and Griffin, W. *Using DNS to Securely Publish Secure Shell (SSH) Key Fingerprints.* IETF, January 2006. RFC 4255.

123. Faltstrom, P., and Kolkman, O. *The Uniform Resource Identifier (URI) DNS Resource Record.* IETF, June 2015. RFC 7553.

124. ICANN Security and Stability Advisory Committee. *SSAC Advisory on Fast Flux Hosting and DNS.* ICANN, March 2008.

125. Klensin, J. *Simple Mail Transfer Protocol.* IETF, October 2008. RFC 5321.

126. Resnick, P., Ed. *Internet Message Format.* IETF, October 2008. RFC 5322.

127. Kucherawy, M. *Resolution of the Sender Policy Framework (SPF) and Sender ID Experiments.* IETF, July 2012. RFC 6686.

INDEX

DNS Security Management, First Edition. Michael Dooley and Timothy Rooney.
© 2017 by The Institute of Electrical and Electronic Engineers, Inc. Published 2017 by John Wiley & Sons, Inc.

IEEE Press Series on
Networks and Services Management

The goal of this series is to publish high quality technical reference books and textbooks on network and services management for communications and information technology professional societies, private sector and government organizations as well as research centers and universities around the world. This Series focuses on Fault, Configuration, Accounting, Performance, and Security (FCAPS) management in areas including, but not limited to, telecommunications network and services, technologies and implementations, IP networks and services, and wireless networks and services.

Series Editors:
Thomas Plevyak
Veli Sahin